THE
SPIRAL TUNNELS
AND
THE BIG HILL

Graeme Pole

THE
SPIRAL TUNNELS
AND
THE BIG HILL

A Canadian Railway Adventure

Altitude Publishing Canada Ltd.
Canadian Rockies/Vancouver

Publishing Information

Canadian Cataloguing in Publication Data
Pole, Graeme, 1956–
The Spiral Tunnels and the big hill
ISBN 1-55153-908-X
1. Spiral Tunnels (B.C.)--History. 2. Railroad
tunnels--British Columbia--History. I. Title.
TF238.S65P64 1995   385'.312   C95-910303-1

Made in Western Canada
Printed and bound in Canada
by Friesen Printers, Altona, Manitoba.

Altitude GreenTree Program
Altitude Publishing will plant in Canada twice as
many trees as were used in the manufacturing of
this product.

Project Development	Stephen Hutchings
Concept/Art Direction	Stephen Hutchings
Design	Stephen Hutchings
Assistant Designer	Sandra Davis
Editing/Proofreading	Noeline Bridge
Maps	Catherine Burgess
	Sandra Davis
FPO Scanning	Debra Symes
Electronic Page Layout	Nancy Green
	Alison Barr
Financial Management	Laurie Smith

Front Cover Photo: Westbound Selkirk 5925
   entering the Lower Spiral Tunnel in 1950.
   Photo by Nicholas Morant
Frontispiece: *Engine Near Field,* A.C. Leighton
Back Cover Photo: The roundhouse at Field in
   1905

Metric - Imperial Conversions
1 metre (m) = 3.28 feet
1 kilometre (km) = 0.62 miles
1 kilogram (kg) = 2.205 pounds
1 tonne = 0.9842 UK tons, 1.102 US tons
1 litre (L) = 0.22 UK gallon, 0.264 US gallon
1 hectare (ha) = 0.004 square miles

10 9 8 7

We acknowledge the financial support of the
Government of Canada
through the Book Publishing Industry
Development Program (BPIDP)
for our publishing activities.

Altitude Publishing Canada Ltd.

1500 Railway Avenue
Canmore, Alberta T1W 1P6

The Big Hill, Lucius O'Brien, 1888

Contents

Chapter 1
Dreams and Reality

Constructed principally between 1881 and 1885, the Canadian Pacific Railway (CPR) was the first transcontinental railway completed in Canada. The CPR was the dream of Prime Minister John A. Macdonald. When British Columbia (BC) joined Confederation in 1871, Macdonald had promised that a railway would be completed within a decade, uniting the new province to eastern Canada.

Macdonald's dream acknowledged neither political realities, nor the physical reality of Canadian landscape, particularly that of mountainous BC. In crossing BC, the tracks of the CPR would be obliged to make roller-coaster traverses of three mountain ranges, and

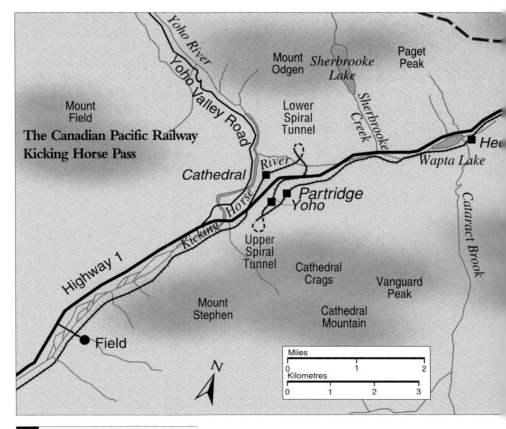

The Canadian Pacific Railway
Kicking Horse Pass

squeeze through the canyons of the Thompson and Fraser rivers while making an end-run around a fourth.

Macdonald had never seen these mountains, with their dense forests, confined canyons, and hazardous avalanche terrain. It is probably just as well. For if he had known about the perils posed by the western slope of Kicking Horse Pass—a place that when tracked with the rails of the CPR would become known as the Big Hill—Macdonald may well have reconsidered, and his dream of uniting Canada by railway steel may never have become reality…

Sir John A. Macdonald, champion of the Canadian Pacific Railway.

"For downright rugged awfulness, there is nothing on the whole of the Canadian Pacific Railway to equal the Kicking Horse Pass."

Stuart Cumberland, *The Queen's Highway, From Ocean to Ocean,* 1887

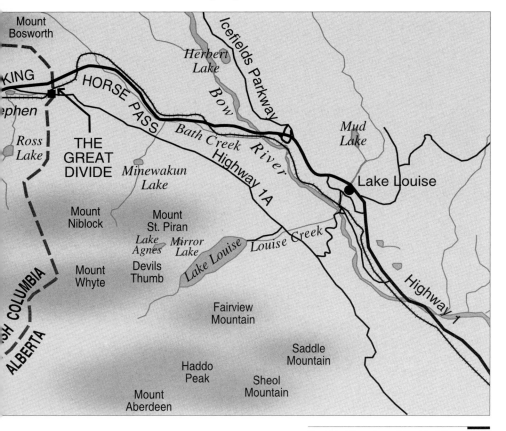

Chapter 2
A Spiral Tunnels Primer

*T*he Canadian Pacific Railway's two Spiral Tunnels are located between the community of Field and Kicking Horse Pass, in Yoho National Park, BC. Completed in 1909, the tunnels reduced the maximum grade on a section of track called the Big Hill, from 4.5 percent (45 m per km, or 238 ft. per mile) to 2.2 percent (22 m per km or 116 ft. per mile).

This grade reduction enabled two eastbound locomotives to haul more than twice the freight up the Big Hill that could previously be hauled by four locomotives, and at five times the speed. Safety was improved for downhill, westbound trains, and scheduling delays, although not eliminated, were reduced.

The Spiral Tunnels are the most popular roadside attraction in Yoho National Park, and can be safely viewed from two viewpoints. However, neither viewpoint gives a complete view of both tunnels, nor a simple understanding of how they create the grade reduction. Thus, their arrangement and operation remain a mystery to most visitors.

How The Spiral Tunnels Work

Eastbound from Field, a train climbs 8 km along a 2.2 percent grade to the lower portal of the Lower Spiral Tunnel in Mt. Ogden. This journey takes approximately 15 minutes.

Tunnel Facts		
	Upper Spiral Tunnel	**Lower Spiral Tunnel**
Location	Cathedral Crags	Mt. Ogden
Length	992 m	891 m
Grade	1.66 percent	1.62 percent
Curvature	288°	226°
Radius	174.7 m	174.7 m
Elevation change	17 m	15.4 m
Viewpoint	km 2.3 on the Yoho Valley Road	Highway 1, 7.4 km east of Field
Length of train required to "loop over itself"	approximately 90 cars	approximately 80 cars

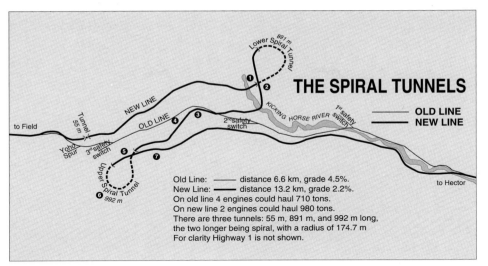

The circled numbers on the map are described in the text below. The route of today's Highway 1 approximates that of the "Old Line."

En route the train passes through the 40-m "Nose Tunnel" in Mt. Stephen, the Mt. Stephen snowshed, and another 55-m tunnel. It also crosses a 69-m bridge. The train then negotiates the 891-m Lower Spiral Tunnel, completing two-thirds of a circle (1). It emerges from the upper portal 15.4 m higher, heading south (2). After crossing the Kicking Horse River again, the track curves southwest. The train thunders by beneath the viewpoint on Highway 1 (3).

West of the viewpoint, the train passes out of view and beneath a highway overpass (4) to the lower portal of the Upper Spiral Tunnel (5). There is 1.6 km of track between the Spiral Tunnels. The train negotiates the 992-m upper tunnel within the lower slopes of Cathedral Crags, completing three-quarters of a circle (6). It emerges eastbound from the upper portal 17 m higher, and once again visible from the viewpoint on Highway 1 (7). The train is now paralleling the original railway grade, surveyed in 1884. Between Cathedral and Partridge sidings, below and above the Spiral Tunnels, the railway line climbs 130 m in a distance of 7.4 km. Westbound trains reverse this sequence.

If placed portal to portal, the Spiral Tunnels would create a shape much like a figure-eight. "Spiral" means that elevation is gained or lost as the trains trace the figure. As many as 30 trains pass through the Spiral Tunnels daily. However, there can be long waits between trains, so please be patient. You can best appreciate the Upper Spiral Tunnel from the viewpoint at km 2.3 on the Yoho Valley Road. The road is generally open from June to mid-October.

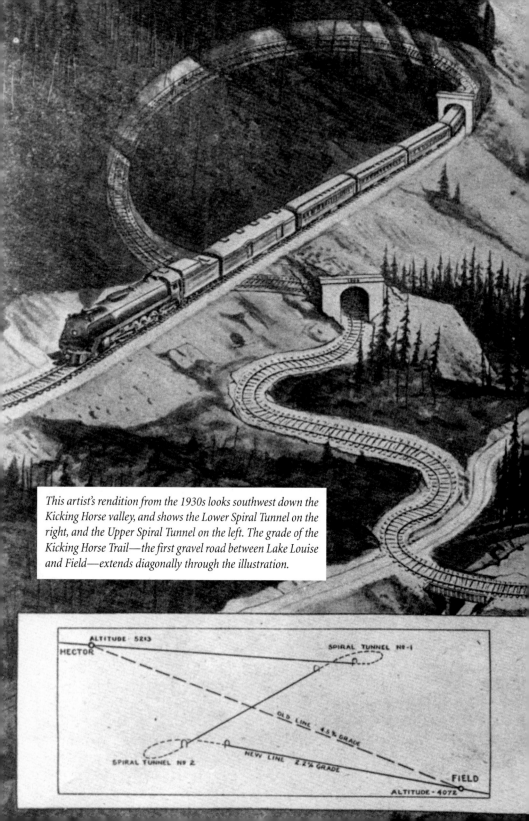

This artist's rendition from the 1930s looks southwest down the Kicking Horse valley, and shows the Lower Spiral Tunnel on the right, and the Upper Spiral Tunnel on the left. The grade of the Kicking Horse Trail—the first gravel road between Lake Louise and Field—extends diagonally through the illustration.

ALTITUDE - 5213

HECTOR

SPIRAL TUNNEL Nº-1

OLD LINE 4.5% GRADE

NEW LINE 2.2% GRADE

SPIRAL TUNNEL Nº 2

FIELD

ALTITUDE - 4072

Headed by an AC 4400 and two SD 40-2 locomotives, CP eastbound extra 9006 negotiates the Lower Spiral tunnel in February 1995. This view is from Partridge Siding, east of the Upper Spiral Tunnel.

Caution!

The Lower Spiral Tunnel Viewpoint on Highway 1 is the most heavily visited viewpoint in any Canadian national al park. More than 3 million people drive past the viewpoint each year, and many of the 700,000 people who annually stop in Yoho include it in their itineraries. Please reduce your speed and obey all highway signs. Do not walk on the highway or its s h o u l - ders.

Chapter 3
From Sea to Shining Sea

vents that led to construction of the Spiral Tunnels began when the province of British Columbia entered Canadian Confederation in July 1871. With a population of only 40,000, the new province's few residents of European origin imagined a bright future that would result from union with the larger nation. The remainder of Canada argued over the wisdom of annexing a vast, scarcely settled, and largely unknown land.

Prominent among the arguments were two in favour: BC possessed abundant natural resources that would aid the growth of the young nation; and many Canadians thought that the distant territory might soon come forcefully under the influence of the United States. The US already had a sizeable population in its western territories and states, and possessed a large standing army as a consequence of the American Civil War.

The Canadian Prime Minister of the day was John A. Macdonald, a Scots Conservative, a visionary powerhouse, and a politician of intellect and commitment. He promoted a "National Policy" that sought to unite the far-flung colonies of British territory in North America. The linchpin in the deal that lured BC into Confederation was Macdonald's promise of a transcontinental railway, linking the new province to eastern Canada. The railway was to be started within two years and completed within a decade. The Liberal opposition party hailed Macdonald's deal as reckless. Some even called it insanity to commit a nation of scarcely 3 million people to the monumental task of financing and constructing a railway across so much uncharted ground.

Through 14 years of turmoil, scandal, and financial conniving, Macdonald championed his cause: the Canadian Pacific Railway. The critics were proved right. The fledgling country was incapable of financing the tremendous work of transcontinental railway building. (The preliminary surveys of the 1870s would cost $3.5 million alone. When construction finally got under way in 1880, it would devour funds at the rate of $1.5 million per month.)

Macdonald organized a syndicate of wealthy British and American businessmen to back the venture and to stem the nation's slide into railway-induced bankruptcy. Not all of Macdonald's dealings were ethical or astute. Amid rightful charges of patronage and kickbacks over the letting of the first railway construction contracts—the Pacific Scandal—his government collapsed in November 1873.

The partially ballasted track of the CPR where it entered the Rockies near Exshaw, Alberta, 1885

Capitalizing on the prevailing anti-syndicate sentiment, the ensuing Liberal government of Alexander Mackenzie attempted to return the financing and control of the railway to Canada. However, with the country's meagre financial resources, Mackenzie failed.

Meanwhile, BC had become restless. The terms of its Confederation deal had not been met. The province talked openly of seceding if work on the railway did not soon begin. This prompted eastern voters to return Macdonald to power in 1878, whereupon he brought his considerable will to bear on the matter. The CPR was finally established as a private venture through an Act of Parliament that received Royal Assent on February 16, 1881.

Mortgaged to the limit, Macdonald's government would eventually bankroll most of the cost of constructing the CPR. When the CPR charter was signed, the railway was given $25 million in cash and title to 10.1 million ha of land along the future line. (The CPR eventually sold most of this land to raise funds, but astutely kept the mineral rights.) In addition, the government picked up the $37 million cost of the railway surveys. In return, the CPR promised to complete the railway within a decade.

After taking possession of existing lines in eastern Canada, the CPR began construction westward on May 2, 1881. The government had started construction of the line eastward from the Pacific coast on April 22, 1880.

A CPR survey crew at Moberly, in the Columbia Valley north of Golden, 1884

While the politicians and businessmen argued, schemed, and dreamed during the 1870s, the wilds of western Canada were host to crews of railway surveyors. These indefatigable souls crisscrossed more than 74,000 km through the Shield country, muskeg, primeval forest, vast prairie, and towering peaks of an unknown land. Almost a quarter of the distance they travelled was subsequently cleared, staked, and surveyed for prospective railway grades. The unforgiving landscape exacted its toll during this labour. By 1877, 38 surveyors had died.

As if their task wasn't difficult enough, the surveyors' efforts were confounded by political bickering about the route that the rails would follow to Pacific tidewater. It was obvious that the western terminus of the CPR would become a major centre, and BC watched developments in this regard carefully. Less obvious to the politicians was the extremely rugged topography of BC and the demands it would make on railway building and operations. To his credit, Sandford Fleming, the first Engineer-in-Chief of the railway, walked across the country to inspect the landscape and the magnitude of the venture first-hand. Fleming's route crossed Yellowhead Pass in what is now Jasper National Park. He thought it feasible, and selected it as the route across the Rockies for the CPR.

Few politicians took Fleming's views to heart. The surveyors remained in the field throughout the 1870s, expending incredible effort exploring routes to which there was

"Only a few feet from the lake [Wapta Lake] it begins its racket of roaring and racing and never lets up until it joins the Columbia." (A description of the Kicking Horse River, from the Edmonton Bulletin, *July 26, 1884)*

no political commitment. In the Rockies alone, six passes were investigated: Kicking Horse, Howse, Athabasca, Yellowhead, Smoky River, and Pine. Curiously, several other likely candidates, including Vermilion Pass, were not seriously considered. Of the passes surveyed, Howse and Yellowhead appeared to offer the most feasible routes. West of the Rockies, the issue was unclear, as no pass had been located through the Selkirk Mountains.

In a startling move, the CPR abandoned the Yellowhead Pass route in January 1882, and settled on the scarcely surveyed Kicking Horse Pass. More than a decade of surveyors' toil was annulled in an instant. The railway moguls and politicians were concerned that branch lines from the northernmost transcontinental railway in the US, the soon-to-be-completed Northern Pacific, would siphon resources away to the south if the CPR followed a more northerly route. They also feared that southern BC and what is now southern Alberta might be claimed by the US if the new railway and the settlement it would bring were not close to the US border.

There were also matters of cost to consider. The proposed route leading to Kicking

Kicking Horse Pass

BEFORE EUROPEAN CONTACT, Natives seldom used Kicking Horse Pass. There is little archaeological evidence of travel or settlement in eastern Yoho National Park. The first party of Europeans to cross the pass, led by two Native guides, was James Hector's outfit from the Palliser Expedition, on September 2, 1858. Four days earlier, near the confluence of the Beaverfoot River and Kicking Horse River above Wapta Falls, Hector had been kicked in the chest by his horse and rendered unconscious. Two days later, his men called the river along which they travelled the "Kicking Horse." Given his injury, and that his party was short of food, it is not surprising that Hector did not report favourably on Kicking Horse Pass as a transportation route.

Hector was a doctor and geologist. Following the Palliser Expedition, he served as head of the Geological Survey in New Zealand, and was knighted. In 1903, he returned to Canada with his son, intending to visit the site of his mishap near Wapta Falls, and to cross Kicking Horse Pass by train. His son became ill at Glacier House, and died soon after. Hector departed immediately for New Zealand, never to return, thwarted in his desire to see the railroad pass that he had helped pioneer.

Horse Pass would be 122 km shorter than a route to Yellowhead Pass. And fewer bridges, and less cutting and filling of the grade would be required on the prairies during construction of a line intended to breach the Rockies at Kicking Horse Pass.

Thus, Kicking Horse Pass was selected, although it is the most rugged pass across the continental divide in the southern Rockies, and the topography immediately west was virtually unknown in 1882. The CPR had chosen a blind alley and was about to stake the company's and the country's financial viability on building a railway across Kicking Horse Pass, and finding corresponding passes for the rails farther west in the Selkirk and Monashee mountains.

It would require a human dynamo to motivate the weary surveyors and wrench a bed for the rails from the inhospitable, trackless bush of mountainous BC. Enter the Major....

Chapter 4
Bush, Beans, and Blasphemy

Although our perception of the 1800s' Wild West is a cliché of colourful characters, Major A.B. Rogers fits the bill. An American, Rogers arrived in Canada in April 1881. He was in his early fifties, yet possessed the spunk and energy of a man half that age. He had become a major by virtue of his service in the Sioux uprising of 1861. A pipsqueak of a man, Rogers wore an outlandish moustache, and possessed an insatiable drive, a profane vocabulary, and a desire for greatness. A near constant stream of expletives issued from his lips, earning him the nicknames: "The Bishop," and "Hell's Bells" Rogers.

In the field, Rogers was as tough as the railway steel that would follow him across the mountains of western Canada. He survived it seemed on a diet of raw beans, hard tack, and chewing tobacco, and expected his men to do likewise. None could suffer under him for more than a few days before complaints, at first culinary, and subsequently regarding matters of humanitarian principle, sprang forth. His men feared "being starved to death or lost in the woods."

Unfortunately, Rogers' spartan outlook, which he thought would save the CPR money and time, eventually cost much more of both in the long run.

Rogers was not originally a "mountain man." He had proven himself as a railway surveyor while staking the grade for the Chicago, Milwaukee, and St. Paul Railroad in the midwestern US. His unlikely selection as pathfinder for the route through the Rockies and Selkirks irked the Canadian surveyors who had spent more than a decade in the field, and drew snide comments from the Canadian press.

However, the CPR realized that the surveying enterprise required new blood. One senior surveyor had written: "There is little probability of a pass being found across the Selkirk range between the upper and lower arms of the Columbia River." Yet with the CPR's new commitment to the Kicking Horse Pass route, this was precisely where the rails would have to be located. In the face of such a firmly entrenched and negative attitude, the CPR was wise to hire an outsider. Rogers possessed the gumption and arrogance to believe that he would succeed where so many others had failed.

Given his personality, it was not likely that Rogers needed more incentive than the honour of being commissioned with what appeared an impossible task. However, James J. Hill, the CPR executive who hired Rogers, sweetened the pot. He promised a $5000 cheque and a gold watch if Rogers succeeded in locating a route through the Rockies and Selkirks that was shorter than the Yellowhead route. Hill also offered to name the pivotal pass through the Selkirks for Rogers, if he should be the first to cross it. With this prospect of glory, Rogers set about his herculean task.

Surveying crews led by Walter Moberly spent three arduous seasons in the 1860s and 1870s, searching high and low for an east-west pass through the Selkirks. They examined all but one major valley on the western slope of those mountains. It was to this valley, the Illecillewaet (ILL-ih-SILL-ah-wet), that Rogers turned in 1881.

Major A.B. "Hell's Bells" Rogers, "The Railway Pathfinder"

Hampered by the deep snowpack of early June, Rogers and a party of 11 surveyors and Natives reached the crest of a pass from the west after an incredibly trying ascent from the present site of Revelstoke on the Columbia River. The pass was a topographic puzzle. It was not oriented east-west as Rogers had hoped, but north-south. In typical fashion, Rogers had not requisitioned enough food, and the hungry party was forced to return without knowing if the pass led across the Selkirks, or into an alpine dead-end.

Following this failure, Rogers travelled to the eastern slope of the Rockies from where he dispatched crews to examine several passes across the continental divide. At a camp in the Bow Valley west of Banff, Rogers met Tom Wilson, who had been hired to pack supplies for the survey. At age 22, Wilson was eager for adventure. In the company of Rogers, he would soon have his fill.

Mt. Stephen, as seen from near where Tom Wilson found Albert Rogers in 1881. The Big Slide is the area in the left centre. Note the footbridge across the Kicking Horse River. The photograph was taken in 1886.

After tersely addressing the crew at the survey camp, Rogers asked among them for a volunteer to serve as his special attendant during the rounds of other work camps. As Wilson later remembered: "Silence greeted his request; there were good reasons for it. Every man present had learned in three days to hate the Major with a real hatred. He had no mercy on his horses or men; he had none on himself. The labourers hated him for the way he drove them and the packers for the way he abused the horses; he never gave their needs a thought. When no one volunteered I thought I might as well take a chance and so took him up."

Wilson accompanied Rogers to Kicking Horse Pass, and in one episode witnessed the impetuous Major take an unplanned bath. (See page 22.) In a later episode, Wilson was privy to a rare glimpse of the human being who dwelled deep within Rogers' inscrutable persona.

Rogers had sent his nephew Albert and two Natives to survey the western slope of the Rockies. They had instructions to cross Kicking Horse Pass from west to east. Only one party of Europeans had done this before—an outfit from the Palliser Expedition of

1858 led by James Hector. Hector's party had found the going very rough, and in his report, Hector described Kicking Horse Pass as "unfit for a carriage road."

Albert was overdue. Concerned for his nephew, the Major instructed Wilson and an assistant to cross Kicking Horse Pass and begin a search. They soon appreciated the tremendous understatement in Hector's remark. It took them a full day of toil through trackless forest and over moss-covered boulders to reach the junction of the Kicking Horse River and Yoho River (today's "Meeting of the Waters")—a distance of less than 10 km. Exhausted, and with no sign of Albert, they pitched camp.

In the evening they were startled by a distant revolver shot. After searching the valley, they found Albert Rogers, near starvation, huddled over a fire.

Next day Wilson and his assistant ascended the pass with Albert in tow and delivered him to the Major. For a moment, the Major's face welled with emotion as he eyed his nephew. Then he slipped back into character with the words: "Well, you did get here, did you? You damn little cuss!", and marched back to camp. Wilson, worn out with the proceedings of the summer, vowed not to return to the Rockies and the Major's survey.

Imagine the torment of Major Rogers as the summer of 1881 ended. The Rockies and Selkirks appeared impenetrable, yet railway steel was advancing unrelentingly toward both mountain ranges. Nonetheless, Rogers kept a brave front when ushered to Montreal to inform the moguls of the CPR of his faith in the Kicking Horse-Selkirk route.

Returning to the mountains with customary haste in May 1882, Rogers led an unsuccessful attempt to reach the crest of the Selkirks from the Columbia Valley to the east. Again, his men were undersupplied, and nearly starved in their struggle with dense forest, torrential rivers, and lingering snow. Temporarily deterred, Rogers embarked again on July 17 from the Columbia River northwest of Golden, and followed the Beaver River to its source in the Selkirks.

For eight days his party struggled through the undergrowth, sometimes covering only 3 km a day. On July 24, they emerged onto the height of the north-south pass that

BC's Many Mountains

THE ROCKY MOUNTAINS are synonymous with BC. Most visitors are surprised to learn that the Rockies are but one range of many in the province. From east to west along the line of the CPR, these ranges are: the Rocky Mountains, comprising the Front Ranges, Main Ranges, and Western Ranges; the Columbia Mountains, comprising the Purcell Mountains, the Selkirk Mountains, and the Monashee Mountains; the Northern Cascade Mountains, and the Coast Mountains. The viability of the route chosen for the CPR hinged on an alignment of suitable passes through the Rockies, Selkirks, and Monashees.

Supplies for the railway survey were often carried on packhorses.

The Major's Bath

WHILE MAJOR ROGERS AND TOM WILSON were searching for Albert Rogers on the east side of Kicking Horse Pass in July 1881, the irascible Major nearly paid a dear price for his short temper. On horseback, Rogers and Wilson came to the edge of a turbulent glacial stream. Although he had only been in the Rockies one summer, Wilson knew that the power of such streams should not be underestimated. He suggested that they camp on the near bank and cross in the morning, when the stream's volume would have dropped.

"Afraid of it are you? Want the old man to show you how to ford it?" And with that, the Major and his steed plunged into the icy waters. The water bowled the horse over in an instant, and Wilson was obliged to rescue the Major, whose temper had understandably cooled. The horse made its own way out of the creek, fortunately on the near bank.

And so Noore's Creek, a principal tributary of the Bow River, was renamed Bath Creek. Thereafter, whenever Bath Creek was in flood and murky with glacial runoff, surveyors and railway workers would sardonically remark, "The old man is taking another bath."

Tom Wilson discovered Lake Louise in 1882. This photograph was taken on the shore of the lake approximately 40 years later.

Rogers had reached the previous summer. Although the route was far from easy, the rampart of the Selkirks had been breached, and Rogers was assured fame. From that day, the clearing where he stood became known as Rogers Pass. Now the Major was obliged to return to the Rockies to complete the survey of Kicking Horse Pass and the intervening ground along the Columbia River.

On his return, the first person Rogers met in the Rockies was Tom Wilson. The Major exclaimed: "I knew you'd be back. You'll never leave these mountains again as long as you live. They've got you now." History would prove the Major right. Wilson had discovered Lake Louise the previous day, and in travels later that summer would make the first recorded visit to Emerald Lake. These two events, continually embellished by Wilson, would promote him to the status of living legend during the heyday of exploration that followed completion of the CPR.

The Major had another mission for Wilson. Although the CPR was committed to the Kicking Horse Pass route, Rogers harboured private doubts. He knew that nearby Howse Pass was scarcely explored. Offering a $50 bonus, and telling him he should be able "to do it in 10 days easy," Rogers instructed Wilson to head north over Bow Pass, follow the Howse River southwest to its source on Howse Pass, and descend the

The $5000 Cheque

As promised, executives of the CPR presented Rogers with a gold watch and a $5000 cheque "as a token of their debt to him for the discovery of the pass for the railway through the Selkirk Mountains, and of their appreciation for his services as Engineer-in-Chief of the location of the mountain section of the railway." Rogers framed the $5000 cheque and was reluctant to cash it. He finally did at the request of the CPR, so that the company could balance its books.

Following the completion of his work in Canada, Rogers took his hard-won reputation as "railway pathfinder" back to the US, to survey the line for the Great Northern Railway. In 1889, while occupied in this work, he fell from his horse. He died soon after from the injuries. Another Rogers Pass in Montana commemorates him.

Blaeberry River to its junction with the Columbia River north of Golden. Virtually nothing was known by either man of the country to be traversed.

The horrible journey required 12 days and nearly cost Wilson his life. When he emerged trail-weary and starving on the bank of the Columbia River, the Major could only snap: "What kept you so long?" From Wilson's description, the Major surmised that the Kicking Horse route was the lesser of two evils. He redoubled his surveying efforts at Kicking Horse Pass.

If Wilson had been better supplied for his exploration of Howse Pass, and not so preoccupied with survival, he would have been able to report more accurately on the route. Although 45 km longer than the Kicking Horse route, the grades on either side of Howse Pass are moderate by comparison, and might have been more suitable for railway construction and operation. By undersupplying Wilson and saving the CPR a few dollars during the survey, Major Rogers helped create a legacy of perpetual expense for the railway in operating the Kicking Horse Pass route.

Rogers completed his survey in November 1883. Only then could he report with absolute confidence to the CPR that a satisfactory route existed across the Rockies and the Selkirks. This eleventh hour news must have been greeted with great relief by the railway's executives, for at that time the "end of steel" was at the entrance to Kicking Horse Pass!

Chapter 5
The Ablest Railway General

ajor Rogers' initial failures in the Rockies and Selkirks in 1881 were only two of many problems that confronted the fledgling CPR. Construction in eastern and central Canada was proceeding at a snail's pace – only 208 km of track had been laid north of Lake Superior. The eastern Construction Superintendent, Alpheus Stickney, was proving inept, and to the CPR's dismay, had embroiled himself in a land sales scandal along the proposed line of the railway in Manitoba.

CPR executive James J. Hill lost no time in rectifying matters. Stickney "resigned," and as his replacement, Hill hired a man touted as "the ablest railway general in the world." It was a grand appellation for a 38-year-old from the US, and the Canadian press collectively let out a derisive howl at the notion that another American would be able to solve the CPR's woes.

William Cornelius Van Horne came to Canada at the climax of a meteoric rise through the ranks of US railroading. After quitting school at age 14 in 1857, Van Horne began work as a telegraph operator with the Illinois Central Railroad. During the next two decades he worked virtually every railroading job on a number of lesser lines, to become General Manager of the Chicago, Milwaukee, and St. Paul Railroad.

Van Horne was a powerful, imposing, direct, and dynamic man. In his quest to master railroading he rarely took a day of rest, and slept little. He was vigilant, meticulous, and renowned for his ability to solve problems, to cut costs, and to increase profits. Van Horne would need these skills and more to tackle the colossal, underfunded venture of the CPR.

Known for his fondness of drink, good food, and fine cigars, Van Horne was also a gambler. In the job offer to oversee the CPR's completion, he must have sensed the incredible risk, and the potential payoffs of power and fame.

From a construction point of view, it was clear that the two most difficult sections of the CPR would be the Shield country of northern Ontario, and the mountain barrier of BC. While deliberating the offer, Van Horne visited the proposed route of the rails north of Lake Superior, and met briefly with Major Rogers. Van Horne had confidence in Rogers, and although one section of the route in northern Ontario seemed to offer "200 miles of engineering impossibilities," he accepted the position as CPR General Manager on November 1, 1881. His annual salary was a princely $15,000.

William Cornelius Van Horne, General Manager of the CPR during its construction

To this point the CPR had been fraught with financial and technical troubles, scandal, and poor public image. The Canadian press was predicting its demise in short order. Now another American was being imported into a position of importance. Whose railway was this?

Except for the executives of the CPR, everyone underestimated Van Horne. In a matter of four years, the "ablest railway general in the world" would oversee the CPR from chaos to completion, and pioneer its rise to the top of the Canadian corporate pile. Always looking for ways to increase revenue, it was Van Horne who established the CPR's telegraph network, hotel chain, and fleet of ocean vessels. These lucrative spin-off enterprises helped assure the railway's immediate and longterm economic survival.

Chapter 6
The Temporary Solution

he CPR's charter stipulated that in no place could the grade of the railway exceed 2.2 percent (22 m per km, or 116 ft. per mile). Major Rogers' surveyed line from Laggan (Lake Louise) to the crest of Kicking Horse Pass followed the Bow River and Bath Creek, at a maximum grade of 1.8 percent.

However, rivers on the western slope of the Rockies characteristically carve much steeper gradients than those on the eastern slope. From the outlet of Wapta Lake, 5.7 km west of Kicking Horse Pass, the waters of the Kicking Horse River plunge westward. In 6.6 km the river bed drops 300 m, creating an average grade in the upper Kicking Horse Valley of 4.5 percent.

To honour the terms of the CPR charter, Major Rogers staked a line for the rails that traversed the southern flank of the Kicking Horse Valley. This created a 2.2 percent grade from Wapta Lake to the mouth of Porcupine Creek, 18.6 km west of the present site of Field.

Hindsight indicates the Major's line would have been a more cost-efficient route. However, the CPR balked at constructing it. To follow the Major's line would have required a 427-m tunnel through Mt. Stephen, and the crossing of many avalanche paths and unstable areas. The CPR estimated the construction costs at $77,500 per km. The work would have required an extra year, and the railway might have gone bankrupt with the resulting delay in commercial operation.

After reassessing Rogers' surveying work, Van Horne put forward a "temporary solution" to the exorbitant cost of the line through the Rockies. He proposed to run the rails straight to the floor of the Kicking Horse Valley from Wapta Lake. The resulting 4.5 percent grade would be more than double the maximum allowed. Van Horne's logic was driven by the need of the moment: complete the railway and open it for business, to generate revenue and prevent the financial collapse of the whole venture.

He also realized that the extensive flats along the Kicking Horse River would make an excellent base for railway operations on the steep hill. Had Major Rogers' line been followed, these flats would have been bypassed and the town of Field as we know it probably would have folded after the railway construction gangs moved on.

Van Horne was most persuasive in lobbying the government for the required change to the CPR's charter. He pointed out that railways in the western US incorporated sim-

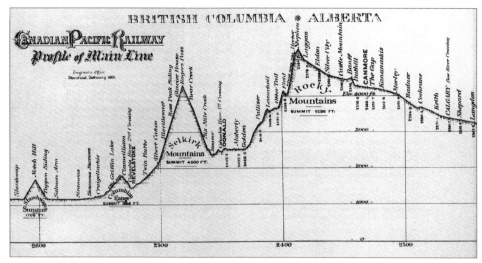

Schematic diagram of elevations on the CPR through the Rockies and Selkirks. Stephen Siding in Kicking Horse Pass (elevation 1625 m/5332 ft) is the highest point on any railway in Canada.

ilar grades. (The Denver and Rio Grande operated an 11 km section at a 4 percent grade.) He stressed that the "temporary" nature of the steep grade would be rectified as soon as time and finances permitted. The charter was hastily amended, and the construction crews were given consent to proceed. Although Van Horne's temporary solution saved almost $2 million in construction costs in 1884, the extra expense of operating regular traffic on the resulting grade soon eclipsed those savings.

On December 6, 1883, the "end of steel" reached the eastern entrance to Kicking Horse Pass, and construction crews retired for the winter. Van Horne's crews had been laying track at a breakneck pace for two summers across the prairies and into the Rockies. In 1882, 671 km of track and 28 sidings were built. On a single day in July 1883, 10.3 km of track were completed. At one point, 152 m of rails were set and spiked in less than five minutes.

The construction gang was formidable. It included the Ryan brothers, two world champion spikers who could routinely drive home each railway spike with just two blows from the sledge hammer; and mighty Big Jack, a Swede who could pick up a 10-m-long, 420-kg section of rail and toss it onto a flatbed car without help.

Nonetheless, the CPR met its match on the hill west of Kicking Horse Pass. Between May and December in 1884, a mere 108 km of track would be laid between Kicking Horse Pass and Donald, on the Columbia River north of Golden. The construction of this section of the CPR was the beginning of a battle with gravity, rock, ice, and snow that has continued in the Rockies and the Selkirks for more than a century.

Chapter 7
The Big Hill

In the spring of 1884, word went out across Canada that 12,000 workers were needed to construct the CPR through the Rockies. A newspaper reported:

"Thousands of men arrived from all points of the compass, and once on the ground they remained because it was practically impossible for them to get out of the country." Construction of the CPR attracted more Euro-Canadians and Americans to the Rockies than had ever seen the mountains to that point. Wages were $1.50 per day; board was $4.50 per week. Morley Roberts, one of the workers, recorded his experience on arrival at Kicking Horse Pass in *The Western Avernus.*

"By this time I was absolutely starving, as it was now the third day since I had had a really satisfactory meal...So we were glad when our train stopped and let us alight. We were received by a man who acted as a sort of agent for the company. He got us in a group and read over the list of names furnished him by the conductor of the train, to which about a hundred answered. He then told us we were to go much farther down the pass, and that we should have to walk about 40 miles, and we could get breakfast where we then were for twenty-five cents. I stepped up and wanted to know what those were to do who had no money...Our friend agreed that we couldn't be expected to go without food, and we had our meals on the understanding that the cost was to be deducted from our first pay. We had breakfast and set out on our 40 miles tramp down the Kicking Horse Pass."

Roberts stayed with the railway construction for two summers, and witnessed the transformation that the CPR brought to the wilderness of BC:

"Round me, I saw the primaeval forest torn down, cut and hewed and hacked, pine and cedar and hemlock. Here and there lay piles of ties, and near them, closely stacked, thousands of rails. The brute power of man's organized civilization had fought with Nature and for the time vanquished her. Here lay the trophies of battle."

One of the greatest blights brought by the CPR was fire. Whether started by locomotives, by careless workers, or by intent, forest fires consumed almost every tree along the CPR's right of way in the Rockies between 1882 and 1885. The effect was so undesirable and long-lasting that in the early 1900s, the CPR retouched promotional photographs of the Rockies, adding green branches to the silver and black spars of burnt trees.

MEN WANTED!

A number of Men will be wanted by the undersigned during the grading season this year on west end of CANADIAN PACIFIC RAILWAY. Wages will be

$1.50 PER DAY,
BOARD $4.50 PER WEEK,

During the Summer Months for good, able-bodied, steady men.

Apply on the work at end of track, now near Cypress Hills, about 600 miles west of Winnipeg.

LANGDON, SHEPARD & CO.,
CONTRACTORS.

END OF TRACK.
April 20th 1883

Construction on the Big Hill below the Old Bridge, in 1884. Mt. Stephen is in the background. Note the uneven rails, typical of new track not yet ballasted for regular service.

The "end of steel" on August 1, 1884. Construction was delayed at this point for several weeks while a bridge was built across the Kicking Horse River. The bridge was near the beginning of the 4.5 percent grade for westbound trains. William Cornelius Van Horne committed the CPR to operating this grade as a "temporary solution" to the excessive construction costs posed by the Big Hill. The Old Bridge still stands adjacent to Highway 1, 8.8 km east of Field.

It was not until the 1950s that new growth began to mask the ravages of these fires.

The first task in constructing the railway grade west from Wapta Lake was to rough out a tote road, along which supplies, equipment, and workers could be moved. In the limestone and quartzite chasm of the upper Kicking Horse River, this initial step required dynamite. The spoil from rock blasting was cleared by hand and trucked away in carts for ballast, or dumped into the river. More than 1.3 million m^3 of rock and debris were moved in this fashion during the 1884 season.

Although only one tunnel was constructed on the original grade—the Nose Tunnel in Mt. Stephen—the work was dangerous. In the confines of the canyon, it was hard to find cover during the blasts, and workers carting the spoil were subject to rockfall from above. Dynamite is a very unstable substance, and its indiscriminate use resulted in many accidents. According to newspapers of the day, construction between Kicking Horse Pass and Golden claimed lives at the rate of "about one per week" in the summer of 1884.

Workers trudged to work daily from a construction camp that was located on the present site of Kicking Horse campground. The camp's stone bake oven, built in 1884, still survives near the Walk in the Past trailhead.

After the rail bed was levelled, bridging crews constructed trestles and snowsheds with timbers obtained on-site. Urgency prevailed in 1884. Scarcely were the bridges completed before the ties were placed and the steel laid. The rail used weighed 42 kg per metre.

The tracks were completed across Kicking Horse Pass on May 28, 1884. From that moment, Van Horne's "temporary solution" earned the dread of all, and became known as the Big Hill. The 6.6-km run to the floor of the Kicking Horse Valley was a construction worker's horror and a railroader's nightmare.

The first construction train to crest the brink of the hill in 1884 ran away, careened from a curve, and was flung into Kicking Horse River, killing three. As construction proceeded down the hill, three safety switches were installed, approximately 1.5 km apart. The switches were tended around the clock, and set to divert trains from the main line onto runaway spurs.

When a work train was 300 m from a safety switch, the engineer gave one long whistle blast to alert the switchman. A further four short blasts were required 100 m from the switch, to indicate that the train was under control, and that the switchman should grant access to the main line. If the four short blasts were not given it meant the engineer had lost control. In this case, or if the train appeared out of control to the switchman, the

The Golden Stairs

A SECTION OF TOTE ROAD in Kicking Horse canyon became known as "the Golden Stairs." Etched into the canyon walls above the turbulent Kicking Horse River, this narrow path was the bane of all. Sandford Fleming, former Engineer-in-Chief of the CPR and a seasoned traveller, descended the Golden Stairs in 1883, and commented: "I do not think I can ever forget that terrible walk; it was the greatest trial I ever experienced."

While descending the tote road with a pack horse, a surveyor checking Major Rogers' work met two men with another pack horse travelling uphill. To make room in that perilous place, one of the horses was pushed over the cliff.

Herbert Samuel Holt was a construction contractor on the mountain section of the CPR, for whom the original siding at Laggan was optimistically named "Holt City." Holt slipped and fell on the Golden Stairs in 1883. He escaped death when he landed on a tree over the abyss. Holt subsequently became a business magnate, president of the Royal Bank, and the wealthiest man in Canada. His empire controlled more than 300 companies. The history of Canadian business would have been very different had the Kicking Horse canyon claimed Holt.

Construction of the tote road on the lower part of the Big Hill in 1884. The tote road was described as "perhaps the worst in the world" in an 1884 newspaper article. "When the mud is not axle deep, rocks and stumps lend a hand to make it interesting for the teamsters." Part of today's Walk in the Past trail follows the tote road, although the going is much easier now.

switch to the main line was not thrown. The train was diverted onto the inclined runaway spur where, as one worker stated: "Wrecks could take place without hindering traffic on the main line."

This elaborate safety precaution was subject to failure. Another construction train ran away. A confused switchman thought he heard the four short whistle blasts and threw the #3 switch. Despite frantic braking efforts by the crew, the train overran the switch. Headed for the Nose Tunnel where 60 men were working, the crew leapt from the train. Complete disaster was only averted when the tender (firewood box) derailed, thus dragging the runaway to a destructive halt before it entered the tunnel.

Because of the excessive cost of operating such a steep grade, the wear and tear on crews and equipment, and the constant risk, the Big Hill was described by one CPR executive as "a heavy cross to bear through the years." Another CPR employee referred to the 14 km of track between Stephen Siding and Field as "the shortest division ever operated by a railroad," but added, "no supervisor ever complained of a shortness of work."

In the 23 years and two months that the CPR operated traffic on the Big Hill, there were several catastrophic wrecks and many derailments. However, not a single passenger was killed. This was remarkable, not only because of the dangers posed by the steep grade, but because by today's standards, rail travel was a perilous proposition in the early 1900s. In the 12 months ending June 30, 1907, 70 people were killed, and 349 injured in passenger rail accidents in Canada. Most of the accidents were caused by track that was improperly built or poorly maintained.

Installing cribwork for the CPR grade along the Kicking Horse River east of Golden, 1885

A westbound pusher locomotive reverses past the #1 safety switch in 1898. This switch was located about 600 m west of Sherbrooke Creek. Highway 1 occupies this grade today; the CPR has been routed along the south bank of the Kicking Horse River. The Old Bridge is to the left, and Mt. Field is in the background. The photograph was taken by M.M. Stephens, station master at Field.

Locomotive 5809, a Santa Fe-type 2-10-2, heads an eastbound passenger train leaving Field, ca. 1930. The train is at the beginning of the 8 km-long, 2.2 percent grade that leads to the Lower Spiral Tunnel.

AUTOMOBILES ROUTINELY NEGOTIATE highway grades of 4 to 8 percent. Why was a 4.5 percent grade on a railway so remarkable?

A moving train has incredible momentum, and can run away on a downhill grade of just 1 percent—1 m of elevation lost in 100 m travelled. Steep grades also make locomotives less efficient and more expensive to operate. An uphill-bound locomotive is 33 percent efficient on a 1 percent grade, but only 16 percent efficient on a 2.2 percent grade. Eastbound trains on the 4.5 percent grade of the Big Hill were far less efficient, being excessively heavy with the fuel, water, and sand required.

A 14-car freight train in the late 1880s typically weighed 762 tonnes. Four locomotives, totalling an equal weight, were required to haul it up the Big Hill. On the prairies, one locomotive could haul four times as much weight, at five times the speed. The extra equipment required on the Big Hill increased operating costs and the risk of mechanical failures, hence the likelihood of runaways, death, and destruction.

Westbound travellers on Highway 1 can best appreciate the steepness of the Big Hill at Sherbrooke Creek, 9.6 km east of Field. Here, the highway is following the 4.5 percent grade constructed in 1884. Compare this with the 2.2 percent grade of the modern CPR line, visible 50 m to the south across the Kicking Horse River.

On the Kicking Horse flats 2 km east of Field, the 2.2 percent grade is obvious. The valley floor here is virtually level. Look south across the river to the railway line. The elevation of the railway tracks above the valley floor is the height that has been gained from Field by the 2.2 percent grade.

Locomotive 562 heads a typical passenger train of the early 1900s, near the crest of the Big Hill at the #1 safety switch. The switchman's shack is alongside the third car from the rear. The runaway spur track is obscured by the rear pusher.

EASTBOUND TRAINS of five cars required a locomotive at the front and a pusher in the rear of the Big Hill. The pusher locomotive was coupled ahead of the caboose. A pusher consumed 7 tonnes of coal on the round trip up and down the Big Hill. The fumes and smoke made life unpleasant for the conductor and rear brakeman in the caboose. Unknown to the CPR, the arrangement also created a serious hazard.

If the coupling ahead of the rear pusher locomotive disengaged, or if the lead locomotive lost power, the rear pusher could ram through the cars ahead. Once this lesson had been learned the hard way, fortunately on a freight train, the rear pusher was engaged behind the caboose. When air brakes became the norm, the brakes on the rear pusher were coupled to those on the lead locomotive. Passenger trains of more than five cars were often split to avoid the dangerous necessity of having to add a second pusher, mid-train.

Run No Risks

FROM THE OUTSET, the CPR published regulations for its crews and switchmen, governing the operation of trains descending the Big Hill. Highlights of the 1904 passenger train regulations included:

- All brakes are to be checked, and retaining power for air brakes is to be increased between Stephen and Hector sidings.
- Cars with faulty brakes are to be cut out of the train at Hector Siding.
- Between Hector Siding and the crest of the Big Hill, the maximum speed is 5 miles per hour.
- On the Big Hill, the maximum speed is 8 miles per hour.
- Brakemen are to dismount and inspect the brake action and the rotation of wheels as the train passes. Any car with wheels skidding is to have braking power decreased just enough to allow wheels to turn.
- No trains are allowed to start down the Big Hill when a passenger train is on the hill, and none are to descend when there is a train at Field unless the siding switch east of Field is set to direct runaways onto a clear siding.
- Safety switches are to be set to direct descending trains onto spur lines.
- One long whistle blast is to be given by the engineer 1000 feet east of each safety switch.
- Four short blasts are to be given at each switchboard if the train is under control, indicating the engineer wants the main line.
- Switchmen must disregard the engineer's request for the main line if the locomotive appears to be out of control.
- Other switchmen and the Field station are to be alerted to runaway trains by sounding an electric gong alarm.
- The journey from Hector Siding to Field is to take 47 minutes.

Regulations for freight trains included:

- Braking power between Hector Siding and the crest of the Big Hill must be sufficient to hold the train stationary on a 3 percent grade.
- In daylight, one locomotive is to be used for a gross weight of up to 300 tons; two locomotives are to be used for 300-550 tons.
- At night, one locomotive is to be used for a gross weight of up to 200 tons; two locomotives are to be used for 200-400 tons.
- The journey from Hector Siding to Field is to take 47 minutes for a train of 5 cars or less; 62 minutes for a train of 5-25 cars.
- The maximum speed allowed on the Big Hill is 6 miles per hour.

The regulations concluded with a stern warning from the Railway Superintendent: "Engineers, trainmen, and switch tenders: Obey the rules, be watchful, and run no risks."

Three locomotives haul a six-car, eastbound passenger train as it passes the #1 safety switch at the crest of the Big Hill. The spur line on the right is for westbound runaway trains. The photograph was taken in 1898 by M.M. Stephens, the Field station master.

A westbound passenger train is stopped on a trestle at the base of the Big Slide, ca. the late 1880s. The Nose Tunnel in Mt. Stephen is just ahead of the train. This "day tunnel" dates to the original construction in 1884. Enlarged in the 1980s, it is still in use.

The Muskeg Summit

Although the Big Hill ended east of Field, the CPR faced another problem created by not following Major Rogers' original survey line. Having descended to the valley floor at Field, westbound trains were obliged to ascend a 2.2 percent grade for 4 km over a rise called Muskeg Summit. From Muskeg Summit, the line descended 5 km to the Ottertail River trestle bridge. The steep grade over Muskeg Summit often required pusher locomotives, and added to the already considerable expense of railway operations near Field.

Construction of the Ottertail Diversion in 1902 eliminated the climb over Muskeg Summit and reduced the grade to 1.5 percent. The Ottertail Diversion routes trains through a rockcut along the bank of the Kicking Horse River at the highway bridge 2.6 km west of Field. The one-way "backroad" from Field follows the pre-Ottertail Diversion railway grade. Farther west on the Ottertail Hill, you can see where the abandoned railway grade winds back and forth, either side of Highway 1.

The 213-m-long Ottertail trestle bridge was dismantled in 1922. An astonishing 295,000 board feet of 12"x 12" timbers was salvaged, along with several tonnes of metal. The earthworks of the trestle are still visible, south of the highway bridge over the Ottertail River.

Figuring the Locomotives

STEAM-POWERED LOCOMOTIVES bore unofficial names such as Jubilee, Pacific, Consolidation, Hudson, Mikado, Selkirk, and Decapod. The names often referred to the railway lines where they were first operated, or to some other descriptive aspect. Locomotives were more formally identified by the configuration of their wheels; for the above, respectively: 4-4-4, 4-6-2, 2-8-0, 4-6-4, 2-8-2, 2-10-4, and 2-10-0.

The first figure indicated the number of unpowered guide or "pony" wheels, the second figure was the number of driving wheels, and the third figure was the number of wheels supporting the rear of the locomotive. (In the Selkirk locomotives, the rear wheels were connected to an auxiliary booster engine.) In all, approximately 3300 steam locomotives, incorporating 23 different wheel configurations, saw use on the CPR between 1881 and 1960. The most in service at one time was 1760, in the 1930s. The average lifespan of a CPR steam locomotive was 26 years, although some, like the Selkirks, were retired decades ahead of their time when diesel-electric locomotives were introduced.

Turning the Locomotives

BECAUSE THEY WERE dedicated to work on the Big Hill, pusher locomotives had to be turned frequently. Turntables were constructed at Field in 1884, and at Laggan. A turntable was a section of track that could be rotated 360°. Next to the turntable was a roundhouse. Locomotives were either turned around to resume service on the Big Hill, or pushed into the bays of the roundhouse for maintenance. Turntables were initially powered by hand. Later they were powered by compressed air from the locomotives, and still later by electricity. The roundhouse at Laggan was removed in 1909. The remains of the roundhouse at Field—probably the third one built there—were removed in 1988 and its foundation was buried in 1993.

The other method for turning locomotives was an arrangement of track called a wye. Together with the adjacent section of siding track, a wye created a slightly embanked triangle. By going forward up one arm of the wye, and reversing down the other arm, a locomotive could make a 180° turn. The earthworks of the wye at Field were removed in 1988. There was also a wye at Stephen Siding, near today's Ross Lake trailhead. It was removed in 1955. However, you can still see some railway ties embedded in the pavement of the 1A Highway. You can see the earthworks of the old wye at Leanchoil Siding from Highway 1, just west of the Wapta Falls road.

Going… Going… Gone

The service record of locomotive 314 illustrated the constant risk that plagued operations on the Big Hill.

Locomotive 314 was a Consolidation 2-8-0, one of nine built in 1885 by the Baldwin Locomotive Works of Philadelphia, especially for work on the Big Hill. On a frosty January night in 1889, 314 was heading a 14-car coal train from Laggan to Field. Behind schedule at Hector Siding, the brakes were eased to try and gain some time. Engineer Jack Spencer managed to keep 314 under control through the first two safety switches. However, by the time 314 approached the third switch, it was a runaway, its wheels skidding and its brakes on fire.

The #3 switchman was new to the job. Awakened by the speeding train, he thought he heard four whistle blasts, indicating that the engineer wanted the main line. The switchman threw the switch and 314 sped past, bound for oblivion. A few moments later, the locomotive derailed into a cliff. The brakeman was killed instantly. The fireman lost both his legs in the wreck and died a few hours later. The engineer, conductor, and a second brakeman survived with minor injuries.

Locomotive 314 was repaired and returned to service. While eastbound from Field in 1894, the water level in the boiler fell too low. The crown sheet failed and the boiler exploded, just west of the Nose Tunnel in Mt. Stephen. The engineer was totally dismembered. The body of the fireman was found far below the track, an equipment handle still in his hand.

The life of 314 was not over yet. Rebuilt once more, its number was changed to 1320 in 1905, and then to 3120. It was scrapped in 1917. You can see part of the dome from the 1894 explosion at the Spiral Tunnel Viewpoint on Highway 1.

82—MOUNT STEPHEN HOUSE FIELD

COPYRIGHT CANADA, 1908
WM. NOTMAN & SON, MONTREAL.

The first hotel in the Rockies: the CPR's Mt Stephen House in 1908. The Field station is on the right in the foreground. Note the recently burned forest on the lower slopes of Mt. Stephen.

Chapter 8
The Canadian Pacific Rockies

he "last spike" of the CPR was driven at Craigellachie (cregg-AL-uh-kih), on November 7, 1885, in the Monashee Mountains, west of Eagle Pass, BC. The CPR was completed five-and-a-half years ahead of its deadline. The first transcontinental passenger train left Montreal on June 28, 1886, and arrived in Port Moody on Pacific tidewater seven days later. Completion of the CPR opened the formidable and formerly impenetrable mountain ranges of BC to all who could afford the fare.

Ironically, it was a problem with operating trains on the Big Hill that propelled the CPR into one of its most lucrative sidelines—the development of a mountain hotel business. As Van Horne had envisioned, the extensive river flats at the bottom of the Big Hill became a logical base for local railway operations. The CPR built a roundhouse and turntable in 1884. The community that sprang up nearby was named after Cyrus Field, promoter of the first trans-Atlantic communications cable, who visited the end-of-steel in 1884.

Two locomotives were required to haul a short passenger train over the Big Hill. The CPR could ill-afford the luxury of including dining cars between Field and Laggan, as this might require another locomotive. Van Horne built a "dining room" next to the Field station platform in 1886. Disembarking visitors were awestruck by the mountain scenery.

Van Horne astutely realized that the facility should be expanded to offer overnight accommodation, and with customary dispatch, saw to it that this was done before the end of the first summer of passenger train operation. Thus was built Mt. Stephen House, the CPR's first mountain hotel. It was designed by Thomas Sorby to reflect a Swiss chalet style. The building underwent annual expansion, with a major addition in 1901 designed by F.M. Rattenbury.

Mt. Stephen House was complemented by Glacier House at Rogers Pass (built in 1887), the Banff Springs Hotel (completed in 1888), and Chalet Lake Louise (built in 1890). Together, these hotels turned the mountains of western Canada into a destination for affluent holiday-seekers, artists, scientists, and mountaineers from Europe and eastern North America. Room rates were from $3 per day.

After a visit to Banff in 1885, Van Horne proclaimed the hot springs there "worth a million dollars," and successfully petitioned the government to establish a park reserve. Mindful of the CPR's desperate need to replenish its coffers, Van Horne did not intend

The Kicking Horse Tea Room, originally called the Big Hill Tea Room, was constructed on the present site of the Lower Spiral Tunnel Viewpoint in 1924. It was torn down in 1954.

The Other Tunnels

YOU MAY BE puzzled by the many openings in the cliffs of Mt. Stephen and Mt. Field, visible from the base of the Big Hill near Kicking Horse campground. These tunnels were not built by the railway, but are the portals of mine shafts and galleries.

Ore was discovered on the slopes of Mt. Stephen in 1882. The Monarch Mine began operation in 1894, extracting lead, zinc, silver, and traces of silica, sulphur, and gold. Hydroelectricity was generated from nearby Monarch Creek, and a 160-m-high aerial tramway was used to transport the ore to the mill in the valley bottom. The Kicking Horse Mine opened in Mt. Field in 1910. The mines changed ownership many times, and operated intermittently until 1952—the last mining enterprises in a Canadian national park. Total production was more than 930,000 tonnes of ore. The mine portals are now gated and locked.

Locomotive 683 heads a royal train carrying the Duke and Duchess of York, as it leaves Field on October 4, 1901. Four pusher locomotives (two more than usual) have been assigned to the ten-car train, perhaps indicating that the CPR did not wish the future King George to come to grief on the Big Hill. The photograph was taken by renowned mountaineer Edward Whymper.

to let other parties easily lay claim to potential tourist dollars in the Rockies.

The same motivation prompted him to request that a reserve be established near Field. The Mt. Stephen Reserve, forerunner of Yoho National Park, was proclaimed on October 10, 1886. Visitors enjoyed the spectacular scenery and made outings to Natural Bridge, and to the fossil beds on Mt. Stephen. Those slightly more adventurous journeyed to Emerald Lake. In an attempt to secure its grasp on the area, the CPR sought to purchase the entire Kicking Horse Valley from the government in 1888. The request was denied.

With convenient access established, the untrodden summits of the Rockies exerted their magnetism on the world's mountaineering elite. The CPR realized the potential for free publicity in the reports the mountaineers would publish, and imported mountaineering guides from Switzerland and Austria to assist them. One mountaineer's report generated great interest in expanding the Mt. Stephen Reserve.

In 1897, a German mountaineer named Jean Habel (AHH-bull) crossed Yoho Pass from Emerald Lake to the Yoho Valley and the foot of a magnificent waterfall. Habel's published account inspired none other than Van Horne to visit the great waterfall. Van Horne named it with the Cree word *Takakkaw* (TAH-kah-kah), which means "it is wonderful!"

At the request of the CPR, the Mt. Stephen Reserve was expanded in 1901. Renamed Yoho, it was to preserve the "glaciers, large waterfalls, and other wonderful and beautiful scenery within its boundaries." *Yoho* is a Cree expression of wonder.

In 1902 the CPR constructed Emerald Lake Chalet. The railway collaborated with the Dominion Parks Branch in developing the Yoho Reserve. Carriage roads were completed to Natural Bridge and Emerald Lake in 1904, to Ottertail in 1905, and to Takakkaw Falls in 1909. National park status was bestowed on the reserve in 1911. By then, Mt. Stephen House had become a world-renowned destination. In the summer of 1912, it registered 8443 guests.

The automobile became common in the Rockies by the early 1920s, and the CPR's monopoly on the tourist trade weakened. Mt. Stephen House, by then obsolete, had been turned over to the YMCA in 1918. To take its place, the CPR constructed automobile "bungalow camps" at Wapta Lake, at Lake O'Hara, and in the Yoho Valley. Together with Emerald Lake Chalet and "tea houses" at Natural Bridge, Yoho Pass, and on the Big Hill, these outlying accommodations catered to the more independent breed of tourist who travelled by automobile.

The glory days had passed. However, the CPR continued to promote tourism and to play a key role in the development of Yoho National Park until it divested itself of tourist accommodation interests between 1935 and 1954. In many cases, the CPR's "auto bungalow camps" in Yoho and elsewhere have evolved into today's commercial accommodation.

A Field in the Mountains

CYRUS FIELD (1819-1892) was an American financier, and proponent of the first trans-Atlantic communications cable. In 1884, Van Horne attempted to entice Field to invest in the CPR, offering free passage to the end-of-steel to see the project firsthand.

At the time of Field's visit, the rails ended at a construction tent camp on the Kicking Horse River flats, west of the Big Hill. Van Horne named the community for the distinguished visitor. Presumably Mr. Field was not sufficiently flattered, for he did not invest in the CPR. However, the town's name endured.

Field is at the highest elevation of any community in BC (1250 m), and is the only CPR community in the Rockies that still bears its original name. Banff was originally called "Siding 29." Lake Louise was formerly "Summit City," then "Holt City," then "Laggan." Golden was optimistically called "Golden City."

Chapter 9
The Maze in the Mountains

While the CPR was developing its mountain hotel chain and advocating the establishment and expansion of national parks, its locomotives continued their labour day and night on the Big Hill. By 1907, there were seven scheduled passenger trains and freight trains daily, comprising an average of 160 cars. With each train requiring two to five locomotives, the air near Field was often thick with smoke. Residents were loath to hang out their clean laundry for fear of its becoming soiled before it was dry. Unburned coal and cinders accumulated at trackside, where in many places you can still see them today.

Events of death and destruction, never commonplace on the Big Hill, had become even less frequent by the early 1900s, but the steep grade and the corporate headache endured. The pusher locomotives each cost $10,000 a year to operate, and required special crews and facilities. Thirty-nine people were employed in the yard at Field alone. With rail traffic increasing (passenger traffic more than tripled between 1901 and 1913), the bottleneck on the Big Hill was becoming extremely costly to the CPR. It was high time to do something about it.

Why had it taken the CPR so long to rectify what one executive had called "this peculiar physical disability?" The original transcontinental line had been completed with such urgency and with so many construction shortcuts that an incredible amount of track and numerous structures needed replacement immediately. Upgrading the line across the country cost $6.2 million in 1886 alone. Consequently, it was two decades before the CPR could muster the finances to reduce the grade east of Field.

The railway considered several options. Other mountain passes in the Rockies were examined in 1902 and 1905, including some that had been ignored 25 years earlier. These passes were rejected because they involved higher elevations, longer routes, or both. The possibility of building a hydroelectric plant on the Kicking Horse River and operating electric trains on the Big Hill was considered and rejected, as was a 16-km tunnel from the crest of Kicking Horse Pass to Field.

There was a proposal to "loop" the rails north into the confines of the Yoho Valley. However, the CPR's costly experience with avalanche terrain in Rogers Pass deterred it from routing the line onto more steep sideslopes. One design engineer suggested that scheduling delays could be solved by simply double-tracking the Big Hill grade.

Locomotive 5809, a Santa Fe-type, leads Selkirk locomotive 5900 on an eastbound passenger train at the Lower Spiral Tunnel, ca. 1940s. The CPR has never operated passenger trains long enough to "loop over" as this one appears to be doing. The train was split to stage the photograph.

The CPR yard at Field, ca. 1931, viewed from the water tower. One of the locomotives in view has been identified as a Selkirk-type, which saw use in the mountains between 1929 and 1952.

Twice the rail and twice the headaches?... Thomas Shaughnessy, CPR President, would have none of that.

After this deliberation, the CPR found a model solution to the problem in the design of the Baischina Gorge tunnels of the St. Gotthard Railway in Switzerland. To avoid steep, straight grades and switchbacks on the Swiss line, their engineers had constructed looping tunnels in which the rails crossed over themselves. CPR Design Engineer J.E. Schwitzer proposed to use two of these tunnels on the Big Hill to double the length of the line from Kicking Horse Pass to Field, and effectively halve the grade.

A 5800 Series 2-10-2 locomotive heads an eastbound passenger train about to enter the Lower Spiral Tunnel, ca. 1927.

The design engineers and moguls of the CPR traded reams of correspondence in honing the plans for Schwitzer's Spiral Tunnels. Matters of cost were foremost. The dimensions of the proposed tunnels were eventually modified to reduce the amount of quarrying.

Finally, the specifications were completed and the construction contract was awarded to the Vancouver company of Macdonnell and Gzowski. Work began in September 1907 and ultimately involved 1000 men. They were paid $2.25 per day.

Crews worked concurrently on both tunnels, advancing headings from either end of each. Some 700,000 kg of dynamite (equivalent to 75 railway carloads) was used to blast 686,000 cubic metres of rock from within Mt. Ogden and Cathedral Crags.

During construction there were five work camps on the Big Hill. The interiors of the tunnels were lit by electricity generated at a hydro power plant built nearby, allowing

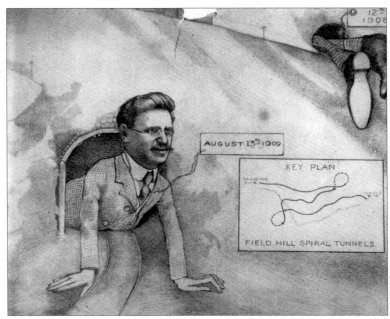

This 1911 cartoon shows J.E. Schwitzer, mastermind of the Spiral Tunnels.

Construction gang at the upper portal of the Lower Spiral Tunnel in 1908.

Workers with a dreadnought drill in the Connaught Tunnel, Rogers Pass, 1915-16. Similar equipment was used to construct the Spiral Tunnels.

work around the clock. The excavators were powered by compressed air delivered from a compressor station through mains 12 cm in diameter. This was the first time compressed air was used to power excavators in Canada.

The excavators loaded construction spoil into dump cars that were hauled from the tunnels by at least two narrow-gauge locomotives that the construction company had purchased for the task. Some of the spoil was used to ballast the grade between the two tunnels. The contract also included construction of a conventional tunnel, 55 m long, located 2.25 km west of the Lower Spiral Tunnel. More than 10 km of above-ground grade reduction was completed on either side of the tunnels, to ensure a maximum grade

The two portals of the Upper Spiral Tunnel during construction in 1908. Note the elevator and footpath to the right of the lower portal. The footpath provided quick access between the grades, and was later used by tunnel patrolmen making their rounds.

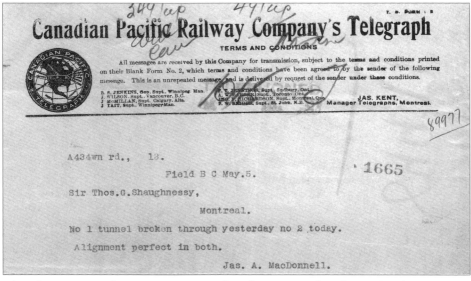

This telegram was sent by contractor James Macdonnell to CPR President Thomas Shaughnessy in 1909, announcing the completion of the Spiral Tunnels. Given the accomplishment, the message is the acme of understatement. Completed in May, the tunnels opened to regular traffic on August 25, 1909. In a July telegram, Macdonnell's partner, C.S. Gzowski, informed Shaughnessy that the work was nearing completion and funds were needed for the payroll. It concluded: "Please do not fail to send some money."

This unique view shows the newly completed Spiral Tunnels and the abandoned Big Hill grade, with track still in place. The buildings are part of the operations of the Monarch Mine in Mt. Stephen, from where this photograph was taken.

of 2.2 percent.

Completed on May 4 and May 5, 1909, the Spiral Tunnels were each 5.2 m wide and 7.6 m high, and had a combined length of 1883 m. They were the largest tunnelling feat undertaken in Canada at the time, and were a remarkable achievement for the days before laser transits and sophisticated surveying techniques. The error of alignment within each tunnel where the crews met was but 5 cm (2 in.). Not surprisingly, Schwitzer's work was viewed favourably by executives of the CPR, and he was promoted to Chief Engineer. Unfortunately, he died of pneumonia soon afterward.

The total cost of the tunnels and grade work was $1.5 million—precisely the amount budgeted. (One-sixth of this cost was for dynamite.) During the 20 months of work, not a single day was lost to labour dispute. The newspapers of the day considered this to be something of a record.

The compliance of the workers was especially surprising given the cost of construction in human terms. At least five workers died. Some were killed when their shovels contacted dynamite that had strayed into construction spoil. Inquests decreed that their deaths were accidental.

FOLLOWING completion of the Spiral Tunnels, the Dominion Parks Branch incorporated part of the abandoned Big Hill grade into a gravel road, for outings from Field. When the road was extended to Lake Louise in 1926, sections of the abandoned grade were again used, and the Old Bridge on the Big Hill was redecked for automobile use. When Highway 1 was completed through Yoho in 1956, it also followed the Big Hill grade.

The Walk in the Past Trail begins in Kicking Horse campground. Pick up an interpretive pamphlet at the trailhead. The 1.2-km trail ascends the route of the railway construction tote road to the CPR main line at Cathedral Siding. It crosses the Big Hill grade, from where the lower portal of the Upper Spiral Tunnel is visible. The trail then follows the grade west along a modern CPR access road to the original #3 runaway spur line and the remains of narrow-gauge locomotive #6, which was used during construction of the Spiral Tunnels.

THE WRECK OF #6 rests beside the grade of a spur from the original CPR main line. For decades, the rusting boiler, frame, and tender were assumed by locals to be the wreck of a runaway pusher. However, railway historians knew that no runaway locomotive had ever been scrapped on the Big Hill.

In the 1950s, the builder's plate was found in the wreckage, identifying it as #7717, a 36" gauge, 2-6-0, built by the Baldwin Locomotive Works of Philadelphia in November 1885. It had been sold to the North Western Coal and Navigation Company (later called the Alberta Railway and Coal Company), which operated between Dunmore, Lethbridge, and the US. To this company, the locomotive was "#6." The locomotive's identity established, the mystery was: Why was it abandoned in the Canadian Rockies, 430 km from its home line, and where no narrow-gauge railway had ever operated?

In 1966, a railway historian realized that the locomotive on the Big Hill in the photo above was also a narrow-gauge 2-6-0. It too had seen service with the Alberta Railway and Coal Company as "#15." When that company's lines were converted to standard gauge between 1903 and 1912, the narrow-gauge stock was sold. Apparently, the construction company of Macdonnell and Gzowski purchased two or more of these locomotives to haul spoil during construction of the Spiral Tunnels. At the end of the job, #15 was probably sold. However, #6 was abandoned, stripped of parts, and left to rust on the Big Hill, where it remains today.

An unidentified pusher locomotive on the original #3 runaway spur beneath Mt. Stephen. The photograph was probably taken in 1885, before regular train service began. The remains of narrow-gauge locomotive #6 are at the end of this spur track. A new #3 runaway spur was constructed 250 m to the east in 1902.

THE SPIRAL TUNNELS have been modified many times. During the original construction, one-quarter of the combined lengths of the tunnels was lined with Douglas fir timbers to prevent rockfall. It was soon discovered that rockfall was a problem throughout, and in 1912 and 1913 the remainder of the tunnels was reinforced.

Large wooden doors were then added to the portals of the Upper Spiral Tunnel. The doors were closed between trains in winter, in an attempt to keep cold air out of the tunnels, thus reducing the build-up of ice from condensed steam. The doors were removed at the end of the steam era in the early 1950s.

Relining and reinforcing the Spiral Tunnels, ca. 1954

By the 1950s, the timber supports in the tunnels were in decay. In 1954, the CPR decided to reinforce the ceilings with steel and concrete. The concrete was sprayed from a high pressure gun onto wire mesh that was hung from the tunnel ceilings, creating a stable surface 45-50 cm thick. Drainage hoses between the mesh diverted seeping water to a trackside ditch. Concrete pillars were erected every 1.6 m to support the new work, which required nine summers and cost $2.5 million.

Until 1992, double-height container traffic could not pass through the Spiral Tunnels. The CPR was losing this aspect of business to the rival Canadian National Railway, whose tracks cross the Rockies at Yellowhead Pass. The CPR considered two options to enlarge the tunnels: lowering the road bed, or shaving and re-shaping the concrete liners to create more headroom. It chose the latter option.

Beginning in April 1992, grout was pumped behind the ceiling liners in the tunnels, and allowed to cure. A machine called a "road-header" then squared the arched ceilings, creating more headroom for the higher cars. The new tunnel liners were secured

A CPR section crew removes ice within one of the Spiral Tunnels, 1973.

to the bedrock with rock bolts. After testing with a special pilot car, double-height freight traffic began to use the Spiral Tunnels in November 1992.

The quartzite and limestone bedrock on the Big Hill contains fissures that channel seeping water. The water freezes into columns of ice in winter. CPR crews used to remove the ice from the Spiral Tunnels to prevent it from interfering with train traffic. In 1988 the CPR began lining the tunnel walls with an insulative closed cell foam that minimizes ice build-up, reducing the amount of ice removal required.

In winter, some freight trains carry an ice rack, mounted on a boxcar just behind the lead locomotives. The rack knocks down icicles that might otherwise damage double-height, piggyback, and automobile cars. The boxcar is ballasted with gravel.

Tie replacement in the Spiral Tunnels has always been difficult, because there is very little working space between the rails and the tunnel walls. In September 1994, the Spiral Tunnels were closed for four days, and concrete ties were installed. The concrete ties are more durable, and should minimize this aspect of maintenance in the tunnels.

No Breathing Room

THE QUESTION OF WHETHER to install costly ventilation vexed the CPR during construction of the Spiral Tunnels. The locomotives of the day were coal-burners. With little traffic, the build-up of fumes and the displacement of oxygen was not expected to create a serious hazard to crews, nor impede the operation of locomotives. The CPR hoped westbound trains would drive any lingering smoke and fumes from the tunnels, and that ventilation would only become necessary if the tunnels were double-tracked.

Nonetheless, J.E. Schwitzer, the engineer in charge of construction of the tunnels, wanted to be certain that the fumes posed no threat. When the tunnels opened on August 25, 1909, Schwitzer rode the first locomotive eastbound from Field to Laggan with all the windows in the cab open. He then boarded the first westbound train to return to Field. Only after riding the second eastbound train through the tunnels was he convinced that they were safe for crews.

Despite Schwitzer's experiment, train crews were exposed to considerable hazard and discomfort in the Spiral Tunnels. At certain times, half a dozen trains might pass through the tunnels within three hours. Then, the build-up of smoke and fumes would become severe, and the temperature could reach 54°C. In an attempt to escape the heat and to reduce the quantity of foul air inhaled, engineers and firemen would sit on the floor of the cab with wet cloths over their mouths as their trains passed through the tunnels.

Engineers were also picky about the make-up of their trains. The engineer would position the cleanest-burning locomotive in front, and ride there. The smokestacks of some locomotives on the mountain section were fitted with deflectors, whose purpose was to keep steam from condensing directly on the tunnels' ceilings and creating ice. However, the deflectors funnelled the smoke towards the engineer's cab, so crews disabled them.

In January 1924, an eastbound freight train stalled in the Upper Spiral Tunnel. The eight crew members were all overcome by fumes, and were treated by a doctor summoned from Field. None of the crew was able to return to work that day, but all were back on duty within a week.

Open-air observation cars were used between Calgary and Revelstoke until the mid-1950s. To prevent passengers from being overcome by fumes, they were ushered into regular cars before the train negotiated the Spiral Tunnels.

Chapter 10
Tall Tales on The Big Hill

Many extraordinary events have taken place in the Spiral Tunnels and on the Big Hill. Some of the railway scenes in the movie *Dr. Zhivago* were filmed there. As one writer of the early 20th century observed, on the Big Hill "it was counted a dull day when something as original as it was startling didn't happen."

Dog Days

A signalman from Field owned a dog that accompanied him on the Big Hill while he travelled on a maintenance speeder. The dog disliked the Spiral Tunnels. When the speeder reached the entrance to each tunnel, the dog would hop off and scamper up or down the bank to await the speeder's emergence from the other portal of the tunnel.

No Encores, Please

The Spiral Tunnels were patrolled hourly by men stationed in houses nearby. Their job was to look for burning timbers, rockfalls, ice, loose rails, and other hazards that might derail a train. During one patrol of the lower tunnel, a worker heard the eerie sounds of piano music. Suspecting the supernatural, he rushed to his section telephone and informed the dispatcher that there was a ghost in the tunnel, and that he would not be completing any more patrols!

The patrolman had been privileged to hear the only "outdoor" concert ever given in the Spiral Tunnels. An employee was being transferred between Yoho and Cathedral sidings on the Big Hill. The last of his possessions to be moved was a piano, which he loaded onto a hand-car. As a crew of section men powered the car and its cargo, the piano's owner could not resist the temptation to tinkle the ivories in the tunnel. What must have been heavenly, reverberant music to his ears, was the fright of a lifetime for the tunnel patrolman.

Look Out Below!

One of the railway sidings on the Big Hill commemorates engineer Seth Partridge. In August 1925, Partridge was working a pusher locomotive east of the Upper Spiral

A steam shovel and work train clear debris from the 1946 jökulhlaup beneath Cathedral Crags.

Tunnel when he heard an avalanche of mud and rocks descending the mountainside above. (Old-timers reported that Partridge claimed to have smelled the slow-moving debris flow.)

Partridge knew that the Yoho Siding house was directly below. He descended the mountainside on foot to warn those slumbering of the imminent peril. Too breathless to yell, Partridge conveyed the urgency of the situation and got everyone outside just before the slide demolished the building. Partridge was honoured many times for his heroic act, including a $1000 award from the American magazine *Liberty*.

As well as destroying the siding house, the Partridge Slide crossed the railway line in three places—the only place on the CPR that this is possible—shutting down operations on the Big Hill. The slide may have been the first recorded instance of the Cathedral jökulhlaup (YOWE-kull-lupp), see p. 71.

A Moose in the Maze

Moose, elk, and deer are commonly seen along the CPR line in Yoho. However, the ungulate experience of one train crew during World War I was far from routine. The engineer of an eastbound freight train noticed a moose, also eastbound, plodding along the tracks toward the entrance of the Lower Spiral Tunnel. No amount of noise from the

"A delightful opportunity for a new sensation..." Passengers on the pilot beam of a westbound passenger train at Field, ca. 1890s.

train's whistle would deter the moose from his, and the train's course. Halfway across the trestle bridge just west of the tunnel, the moose slipped and popped all four legs between the railway ties. Belly-down on the ties, legs flailing, the moose brought the train to a stop.

With great nerve and brawn, the train's crew hoisted the errant moose from the rails and carried it across the bridge, whence it again demonstrated its one-track mind. As the train crew watched in disbelief, the undeterred moose sauntered into the Lower Spiral Tunnel. The train was obliged to slowly follow the animal around the bend to the grade above, where the moose departed.

First Lady on the Cowcatcher

In 1886, Prime Minister John A. Macdonald and his wife, Agnes, crossed Canada by rail. For Macdonald it was an opportunity to gain a better understanding of the country that the CPR was helping to unite. Weary from the 14 years of bitter struggle that completion of the railway had entailed, Macdonald was a recluse for much of the trip. Lady Macdonald was more inclined to venture forth. At Laggan Siding, she created a stir.

While pusher locomotives were engaged for the descent of the Big Hill, Lady

Locomotive 1602 on the turntable at Field in the early 1900s. Built in 1904, it had a 2-8-0 wheel configuration, and saw service until 1951.

Macdonald announced that she would ride on the foremost pilot beam (cowcatcher) "from summit to sea." She reasoned it would provide "a delightful opportunity for a new sensation." Although the Prime Minister did not approve, his wife reportedly made good on her word, and in those days before liability concerns were rampant, riding the pilot beam became the rage.

The One That Got Away

The Big Hill and the flats east of Field are sometimes swept by a fierce northeast winter wind, known to locals as a "Yoho Blow." In the late 1800s and early 1900s snowfalls were more substantial than in recent decades. (Annual snowfalls in Kicking Horse Pass were 8.35 m at the turn of the century. The 22-year average, ending in 1992, was 4.81 m.) The high winds of a Yoho Blow piled immense drifts on the Big Hill, often taller than a locomotive's smokestack.

The track was cleared with rotary plows and a massive, 5-metre-wide wedge plow. One day, a wedge plow was on the front of a train, eastbound from Field. When the train emerged from the Lower Spiral Tunnel, the engineer noticed that something was wrong. The 41-tonne plow was gone, and so was its crew!

Selkirk locomotive 5929 and Santa Fe-type locomotive 5813 head an eastbound passenger train Partridge Siding, east of the Upper Spiral Tunnel, ca. 1945.

In disbelief, the engineer backed his train down the hill to the telegraph office at Field, where spotters piled aboard. The train started back up the hill. At the Lower Spiral Tunnel, they found the plowman cursing his way to trackside through the snowy depths. He informed the engineer that the plow and cab now rested at the river's edge, 100 m below, and the miraculously uninjured crew would be most happy if someone would come and dig them out!

Rush Delivery

FINDING HIMSELF AT THE controls of a runaway freight on the Big Hill, a brash engineer proclaimed to his mortified fireman, "Here goes for Field!" The engineer gave the four whistle blasts at each safety switch, indicating that he was in control and that he wanted the main line. In disbelief, all the switchmen complied, although the train appeared bound for destruction.

When the engineer brought the train safely into Field, he basked in the glory of being the first to pilot a runaway the length of the Big Hill. But briefly. He was also the last. The station master handed the engineer a telegram that indicated he was fired for his flagrant disregard of the rules. The telegram was derisively marked "Rush."

Selkirk locomotive 5904 on the turntable at Field. Built in 1929, 5904 and the 36 other Selkirks were the zenith of steam power. Following the arrival of diesel locomotives in the mountains in the 1950s, all but two of the Selkirks were scrapped.

Chapter 11
The Big Hill Today

*T*he Big Hill has witnessed an astounding evolution of railway locomotives and rolling stock. When the CPR was completed, locomotives were fired with firewood. With a 4-4-0 wheel configuration, they typically produced a tractive effort of 6100 kg. The coal-burning locomotives that followed, such as the 2-8-0 Consolidations, were more powerful, generating a tractive effort of 10,900 kg. In 1912, the CPR began converting some of its steam locomotives in the mountains from coal to oil, and typical tractive effort achieved was 17,200 kg.

The Selkirk-type locomotives, of which 37 were built for the CPR between 1929 and 1949, were the monarchs of steam power in the CPR's mountain operations. They had a 2-10-4 wheel configuration: two unpowered guiding wheels, ten driving wheels (each 1.6 m in diameter), and four trailing wheels that were coupled to an auxiliary booster engine. The booster was used to assist in hill starts, and could provide additional power for speeds up to 19 km per hour. You can readily identify the Selkirks in photographs. They bore numbers 5900 to 5935. An experimental Selkirk bore number 8000.

Weighing 371 tonnes, a fully loaded Selkirk and tender carried 18,655 L of bunker-C fuel oil, and 54,600 L of water. These monsters of the steam era generated 35,000 kg of tractive effort, with another 5440 kg available from the booster. On a flat grade, a Selkirk could haul 2805 tonnes—45 passenger cars. Four Selkirks were required to power a 4064-tonne eastbound freight train up the Big Hill, but a single Selkirk could manage a 16-car, 1000-tonne passenger train from Calgary to just east of Rogers Pass, without pushers. Accordingly, individual Selkirk locomotives saw service as pushers, or as "through-trains."

The Passing of Two Eras

RAILWAY ENTHUSIASTS lament the passing of two traditional aspects of railroading on the Big Hill. Regular passenger train service through Kicking Horse Pass was discontinued in January 1990. The thrill of riding through the Spiral Tunnels is now only available to passengers on a tour train, the *Rocky Mountaineer,* that operates between Vancouver and Calgary in the summer. In 1993, there was talk of resurrecting regular passenger service, but the start-up cost was estimated at $25 million. Cabooses were removed from regular freight service in 1990, and are now rarely seen.

Locomotive 732, a Consolidation-type 2-8-0, on the turntable at Field. Built in 1899, it saw service until 1929.

Selkirk locomotive 5928 heads an eastbound passenger train as it crests the Great Divide at Stephen Siding, ca. 1940s. This train probably had a pusher cut-out at Stephen.

Each Selkirk consumed about 250 L of water and 113 L of oil for each kilometre travelled in the mountains. The Selkirks and other locomotives also carried large amounts of sand, which was applied to the tracks to create traction for climbs, and friction for braking.

Despite their prowess, the Selkirks were usurped from mountain service by diesel-electric locomotives in 1952. The Selkirks saw further use on the prairies until 1959, by which time most had been scrapped, despite years of working life that remained. There are now only two Selkirk T1c locomotives in existence: #5934 is at Heritage Park in Calgary, and #5935 is in the Canadian Railway Museum, in Delson, Quebec.

The CPR put its first diesel-electric locomotive in service in 1937, but it was 1952 before they were used on the Big Hill. Today's SD40-2 and AC-4400 diesel-electric locomotives weigh 250 tonnes, generate 3000 horsepower, achieve a tractive effort of 32,200 kg, are capable of 105 km per hour, and have a price tag of $2 million. The tremendous power of these locomotives permits long freight trains—110-car, sulphur, or grain trains are common. The value of locomotives and rolling stock on such a train exceeds $20 million.

With a cargo of 13,200 tonnes, a 1.6-km-long, 110 car freight train on the Big Hill is typically powered by six diesel-electric locomotives—four out front, and two additional computerized "slave" locomotives, mid-train. On-board computers coordinate the acceleration and braking between the lead and slave locomotives. This is necessary because when a train crosses Kicking Horse Pass, the lead locomotives can be on a downhill grade, while the slaves are still going uphill.

The freight cars of most eastbound trains on the Big Hill are empty, and the trains are relatively lightly powered, sometimes with only two lead locomotives and no

The Lake Louise Grade Reduction

AS YOU DRIVE along Bath Creek between Lake Louise and Kicking Horse Pass, you may not think the valley floor is steep. However, with a grade of 1.8 percent, this section of track was costly for the CPR to operate. In 1980, the railway completed a $13.9 million grade reduction that entailed double-tracking.

Westbound trains are generally loaded as they cross Kicking Horse Pass, whereas eastbound trains are often empty. A new westbound track was constructed, with bridges, cuts, and fills that uniformly spread the 100-m climb from Lake Louise to Kicking Horse Pass, creating a 1 percent maximum grade over the distance. Eastbound trains follow the steeper, old track. You can see the double-tracking from several places on Highway 1 between the Icefields Parkway junction and Kicking Horse Pass. The westbound track crosses Highway 1 on an overpass, just west of the Icefields Parkway junction.

Workers clear a snow avalanche from the tracks east of The Big Slide in the early 1900s.

The Lake Louise section crew with a wedge plow at Lake Louise station in 1978. This plow dates to the 1930s. You will often see it in the Field yard.

mid-train slaves. However, loaded eastbound lumber trains may have as many as eight locomotives out front, with no slaves mid-train.

The factor limiting the size of today's freight trains on the Big Hill is not a lack of power available to climb the grade, nor insufficient braking power to descend it. Cathedral Siding and Yoho Siding on the Big Hill cannot handle extremely long trains, which must be specially scheduled. Special scheduling adds to the already complicated process of coordinating the movements of eastbound and westbound trains along the single track. So it is avoided unless necessary, and shorter trains prevail.

A Snow Job

Despite tremendous improvements in technology and in the efficiency of modern locomotives, the CPR must still battle the elements on the Big Hill.

Snow avalanches are the principal scourge of railways in the mountains. Snowsheds and tunnels counter the threat. The scope of the avalanche problem on the Big Hill is not as severe as at Rogers Pass. However, the major slides that occur cannot be controlled by anything except a tunnel. Since the original estimate of $5 million in 1907, the cost and logistics of putting the entire grade east of Field underground have become prohibitive.

This eastbound locomotive pulled into the Field yard in February 1990 plastered in snow. The engineer had plowed through avalanche debris on the tracks.

Trains are protected from running into the debris of small slides by avalanche fences. Installed on the uphill side of the track, a fence's electrical circuit is broken when a slide sweeps through. This illuminates warning signals at trackside and at the CPR's Centralized Traffic Control in Calgary, advising trains to stop.

The Big Hill grade passes through the runout zones of three enormous avalanche paths. The most westerly of these, known as the Big Slide, is at the bottom of the Big Hill, beneath a broad gully on Mt. Stephen. This avalanche path is more than a kilometre long. Most avalanches here are triggered when ice calves from the terminus of the glacier above, and lands on the snow-covered slope. By the time the billowing mass of ice and snow reaches the railway and highway, it usually has incorporated rock debris. Thus, there is considerable potential for destruction.

Avalanches at the Big Slide have been as common as they have been catastrophic. Over the years, the CPR installed a series of wooden sheds to deflect the avalanches, and trestles to elevate the tracks above the sliding snow. However, large slides would still inundate the track.

The last straw for the CPR was in 1986 when an ice avalanche derailed a sulphur freight train. The damage to railway cars, the costly environmental clean-up, and the poignant reminder of what might have happened had the freight been volatile gas or liquid, prompted the CPR to build a concrete snowshed, completed in 1988.

During construction of the shed, D-9 Cats sculpted the gully. The intent of the landscaping was to divert avalanches away from open track, and over the roof of the shed. On

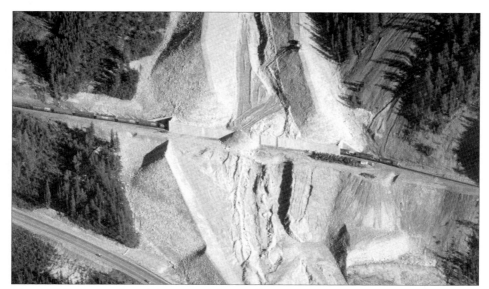

The Big Slide: Photographer Edwin Knox scaled Mt. Field to take this photograph two days after the August 1988 slide.

the evening of August 25, 1988 a torrential downpour struck Yoho. The gully funnelled water toward the concrete snowshed. The banks of the gully were undercut by the deluge, and the material disturbed by the CPR's recent contouring efforts collapsed. Suspended in the water, the scree, gravel, and mud created a debris flow with enormous power.

The concrete snowshed acted as a dam, temporarily impeding the debris flow until it both overtopped the shed and undermined the bed of the railway. With the resulting surge, the flow shot into the Kicking Horse River with disastrous results. The river was temporarily diverted, flooding Highway 1 and Kicking Horse campground. The highway was closed for 20 hours.

Another deluge on August 3, 1994 produced similar results. The CPR's engineering marvel may have defeated the Big Slide, but it has created a new opponent—the Big Slurry!

Slightly west of the Spiral Tunnel Viewpoint on Highway 1, the tracks cross the CPR's second major adversary on the Big Hill. High up the mountainside to the south is a notch between the towers of Cathedral Crags and the unnamed northern outlier of Cathedral Mountain. Early in the 20th century, a lobe of Cathedral Glacier flowed through this notch. Today the glacier has receded. However, its meltwater occasionally discharges violently through the notch in a phenomenon called a jökulhlaup (YOWE-kull-lupp). This Icelandic word means "glacier flood."

The water drains a lake situated beneath the glacier. When the glacial melt exceeds the volume of the lake's rocky basin, the water spills out from beneath the ice, pours through the notch, and sweeps the gully to the railway and highway below. The Partridge Slide of 1925 (see p. 59) was certainly a jökulhlaup. There were jökulhlaups in 1946, 1954, 1978, and 1984, and archival photographic evidence suggests at least one jökulhlaup around 1909. The 1978 event deposited material 6 m deep at Yoho Siding, and 2 m deep on Highway 1. The CPR line was closed for two days.

In 1985, the CPR installed pumps on Cathedral Glacier, more than 1300 m above the railway. When the lake's level rises in late summer, the pumps divert the excess water harmlessly over the west face of Cathedral Mountain into Monarch Creek. There has not been a jökulhlaup since the pumps were installed, although during heavy rains in the summers of 1993 and 1995, the pumping efforts only just managed to keep ahead of the accumulating meltwater. The jökulhlaup path is clearly visible from the Upper Spiral Tunnel Viewpoint, at km 2.3 on the Yoho Valley Road.

The CPR's last major concerns on the Big Hill are the Mt. Bosworth slide paths, 1.4 km west of Kicking Horse Pass. In 1972 and 1990, avalanche control work on these paths released slides that buried the CPR line and Highway 1. You can still see dead trees from the 1972 avalanche on the shores of Sink Lake.

Although some people (and many CPR employees) may curse nature's fickle and disruptive ways, and the havoc that avalanches, debris flows, and rockfalls create on the Big Hill, many people find assurance that technology cannot master nature in this corner of the Canadian Rockies.

Sober Reminders

THE BIG HILL'S SAFE REPUTATION was tarnished by a series of incidents in the 1990s. In April 1994, fourteen cars of a 110-car potash train derailed in the Upper Spiral Tunnel, forcing closure of the track for six days. In April 1996, a potash train derailed at the switch immediately east of Field. On December 2, 1997, the Big Hill saw its first true runaway in more than fifty years. CP westbound Extra 9558, with an 88-car grain cargo, lost braking power at the crest of the hill. Nineteen cars derailed in the Upper Spiral Tunnel, damaging the tunnel supports and interior. Another forty-nine cars derailed 2 km east of Field. The track was closed for a record nine days. After investigating the derailment, CP Rail fired the train's engineer for not following procedures. On January 2, 1998, another westbound grain train ran away on the Big Hill, but fortunately remained upright. These events, involving inert cargoes, are sober reminders of the potential for calamity: Seven percent of the cargo carried by CP Rail is toxic, flammable, or hazardous.

Selected Place Names on the CPR

Beaverfoot Surveyor G.M. Dawson reported that this was a translation of the original Stoney name for the river and mountains that now form Yoho's southwest boundary.

Bosworth George Bosworth was a Vice-President and Freight Manager of the CPR when the 2771-m mountain north of Kicking Horse Pass was named for him in 1903.

Burgess Alexander Burgess was a Deputy Minister of the Interior during construction of the CPR. The 2599-m mountain is immediately northwest of Field.

Cathedral Mountain (3189 m) was descriptively named by mountaineer James Outram (OOT-rum) in 1901. The Upper Spiral Tunnel is within the lower slopes of adjacent Cathedral Crags (3073 m), also named by Outram, and first climbed by him in 1900.

Chancellor Sir John Boyd was the Chancellor of Ontario in 1886 when he arbitrated a mineral rights dispute between the CPR and the federal government. Chancellor Peak (3280 m) towers over the Kicking Horse Valley in the southwestern part of Yoho.

Dennis John Stoughton Dennis served as the Surveyor General of Canada, and was a Deputy Minister of the Interior when the CPR was constructed. The 2539-m mountain is immediately south of Field.

Field William Cornelius Van Horne named the settlement for Cyrus Field, promoter of the first trans-Atlantic communications cable, who visited the end-of-steel west of Kicking Horse Pass in 1884. Van Horne was attempting to flatter Cyrus Field in the hope that he would invest in the CPR. He didn't. Mt. Field (2635 m) is immediately north of town.

Hector The siding near the east end of Wapta Lake was named for Sir James Hector, medical doctor and geologist with the Palliser Expedition of 1857-1860. Hector was in the first party of Europeans to cross Kicking Horse Pass, in 1858.

Hunter Explorer James Hector named this 2615-m mountain, possibly for his guide Nimrod, whose name means "great hunter." This was the highest form of flattery, for Nimrod was a poor hunter, at least while in Hector's service.

Hurd Mt. Hurd (2993 m) is immediately southwest of the Ottertail River mouth. It commemorates M.F. Hurd, a CPR surveyor and an assistant to Major Rogers. A painting of Mt. Hurd by renowned artist F.M. Bell-Smith was featured on the 1928 10-cent Canadian postage stamp.

Kicking Horse James Hector of the Palliser Expedition was kicked by his horse near Wapta Falls in 1858. His men gave the name to the river two days later. The pass is at the easternmost headwaters of the river.

Laggan George Stephen and Donald Smith of the CPR executive were Scots, and many sidings on the CPR in the mountains were given Scottish names. The siding name was changed to Lake Louise in 1913.

Narao (Nah-RAY-owe) Derived from the Stoney language, this word reportedly means "hit in the stomach"—possibly in reference to James Hector of the Palliser Expedition, who was kicked by his horse in 1858. The 2974-m peak is southwest of Kicking Horse Pass.

Niblock John Niblock was Superintendent of the western division of the CPR in 1894, when the 2976-m mountain southeast of Kicking Horse Pass was named.

Ogden Isaac Ogden was an auditor and Vice-President of the CPR. The 2695-m mountain that contains the Lower Spiral Tunnel was named for him in 1916, the year after his death.

Ottertail "Ottertail" is a translation of an unrecorded Native name, applied by geologist and explorer George Dawson to features in the valley 10.1 km southwest of Field.

Paget Reverend Dean Paget (PADGE-ett) of Calgary was a founding member of the Alpine Club of Canada. The 2560-m mountain north of Wapta Lake was named for him when he climbed it in 1904.

Partridge This siding commemorates engineer Seth Partridge and his act of bravery in 1925. (See p. 59) Previously it had been known as Mars Siding.

Sherbrooke The creek and nearby lake at the top of the Big Hill were named by surveyor J.J. McArthur for the town of Sherbrooke, Quebec.

Stephen Sir George Stephen was President of the CPR during its construction. The name was applied to the 3199-m mountain southeast of Field in 1886, and to the railway siding in Kicking Horse Pass. After the mountain was named for him, Sir George requested that his name be changed to Lord Mount Stephen. Thus, a man was named for a mountain! You may see Stephen Siding at the Great Divide on Highway 1A, 16.4 km east of Field.

Van Horne William Cornelius Van Horne was General Manager of the CPR during its construction, and subsequently became its second President and Chairman of the Board. The mountain range west of Field was named for him in 1884. Van Horne took Canadian citizenship in 1890, and was subsequently knighted. He constructed the first railway across Cuba in 1902. He died in 1915.

Vaux (Vox) Explorer James Hector named this magnificent and lofty mountain for William Vaux, a curator at the British Museum. The 3319-m peak dominates the lower Kicking Horse Valley.

Wapta is derived from a Stoney word for "river," and was the original name of the Kicking Horse River. The 2778-m mountain is north of Mt. Field. Wapta Lake is 3.7 km west of Kicking Horse Pass. Wapta Falls is in the southwestern extremity of Yoho.

Whyte Sir William Whyte was a Vice-President of the CPR when the 2983-m mountain south of Kicking Horse Pass was named in 1898.

Yoho A Cree expression of awe and wonder.

Recommended Reading

Bain, Donald. *Canadian Pacific in the Rockies.* Calgary: The British Railway Modellers of North America. (In ten volumes.)

Berton, Pierre. *The National Dream,* Toronto: McClelland and Stewart, 1970.

Berton, Pierre. *The Last Spike,* Toronto: McClelland and Stewart, 1971.

Berton, Pierre. *The Great Railway Illustrated.* Toronto: McClelland and Stewart, 1972.

Cruise, D. and Alison Griffiths. *Lords of the Line.* Markham: Penguin Books, 1989.

Garden, J.F. *Nicholas Morant's Canadian Pacific.* Revelstoke: Footprint Publishing, 1991.

Lamb, W. Kaye. *History of the Canadian Pacific Railway.* Toronto: Collier Macmillan, 1977.

Lavallée, Omer. *Van Horne's Road.* Montreal: Railfare Enterprises, 1974.

McKee, Bill and Georgeen Klassen. *Trail of Iron.* Calgary: Glenbow-Alberta Institute, 1983.

Pugsley, E.E. *The Great Kicking Horse Blunder.* Vancouver: Evergreen Press, 1973.

Turner, Robert D. *West of the Great Divide: The Canadian Pacific in British Columbia 1880-1986.* Victoria: Sono Nis Press, 1987.

Yeats, Floyd. *Canadian Pacific's Big Hill.* Calgary: The British Railway Modellers of North America, 1985.

Acknowledgements

THE TEXT WAS reviewed by:

Donald Bain, British Railway Modellers of North America, Calgary; Graham MacDonald, Historical Services, Parks Canada, Calgary; Peter Corley-Smith; Lance Camp; and Robert Turner, Curator Emeritus of the Royal British Columbia Museum, Victoria.

The author is grateful to the following for research assistance:

Marnie Pole; David Mattison and Kelly Nolin, British Columbia Archives and Records Service, Victoria; Stephen Lyons, Canadian Pacific Corporate Archives, Montreal; Ruby Nobbs and Cathy English, Revelstoke Museum and Archives; Colleen Torrence, Golden Museum; John Gaffney and Tom Gough, CPR employees at Field; Donald Bain, British Railway Modellers of North America; Robert Turner, Curator Emeritus of the Royal British Columbia Museum; Dr. Lionel Jackson, Geological Survey of Canada, Vancouver; Clive McKay, CP Rail Limited, Montreal; Carol Haber, City of Vancouver Archives; Donna Cook, Gord Rutherford and Terry Willis, Yoho National Park; the staff of the Glenbow-Alberta Institute, Calgary; the staff of the Whyte Museum of the Canadian Rockies, Banff; Susan Bridgman and Laurie Robertson, Vancouver Public Library.

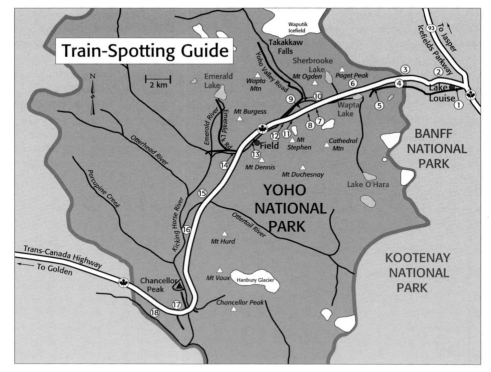

Train-Spotting Guide

2 km

Train-Spotting Guide

The following locations are excellent for train-spotting. Please keep your safety and the safety of other highway users in mind. Park only in pull-offs, or well onto the highway shoulder. Walk facing traffic. The CPR right-of-way, signals and switches, are private property.

1 Lake Louise Station Follow Sentinel Road north to its end in Lake Louise village. Built in 1909, the Lake Louise Station was proclaimed a National Historic Railway Building in 1990. There are eastbound and westbound tracks here, and dining cars restored as part of the Station Restaurant.

2 Bath Creek Highway 1 parallels the CPR for 6.2 km west of the Icefields Parkway junction. You can park safely at several pull-offs. The westbound (upper-level) track was completed in 1980 as part of a $13.9 million grade reduction.

3 Kicking Horse Pass Park at the pull-off on the north side of Highway 1, just east of Kicking Horse Pass. The switch here directs trains between the upper and lower tracks of the Lake Louise Grade Reduction.

4 Stephen Siding Walk 100 m north from the Great Divide on Highway 1A to the east end of the 123-car Stephen Siding. At an elevation of 1625 m, this is the highest point on any railway in Canada. The nearby monument to James Hector was erected in 1906.

5 Hector Crossing Highway 1A crosses the CPR 150 m south of Highway 1. You can park nearby and safely watch trains near the crest of the Big Hill grade.

6 Wapta Lake Park at West Louise Lodge or at the Wapta Lake picnic area, 11 km east of Field. Carefully cross Highway 1 to the footbridge at the outlet of Wapta Lake. The CPR is nearby, and you can see the footings of an 1884 railway bridge. This bridge was abandoned in 1960 when Highway 1 was completed and the CPR was relocated to the south bank of the Kicking Horse River. You can obtain fine distant views of trains from the Wapta Lake picnic area.

❼ Lower Spiral Tunnel Viewpoint Here, at the most famous railway viewpoint in Canada, you can see long trains loop over themselves in the Lower Spiral Tunnel in Mt. Ogden. If you look carefully, you can see the upper portal of the Upper Spiral Tunnel on the slopes of Cathedral Crags, to the southwest.

❽ The 1908 Tunnel Adjacent to Highway 1, 3.2 km west of the Lower Spiral Tunnel, this 55-m "day tunnel" was completed as part of the Spiral Tunnels contract. It is located at the west end of Cathedral Siding. Unfortunately, there is no longer a highway pull-off nearby.

❾ A Walk in the Past This trail begins in Kicking Horse campground. See pages 54-55.

❿ Upper Spiral Tunnel Viewpoint From this viewpoint at km 2.3 on the Yoho Valley Road, you can see the portals of the Upper Spiral Tunnel on the slopes of Cathedral Crags.

⓫ Mt. Stephen Snowshed This structure was completed in 1988 to protect the tracks at the Big Slide. You can best appreciate the snowshed from the parking lot at the beginning of the Yoho Valley Road.

⓬ The Nose Tunnel This 40-m "day tunnel" through the lower cliffs, or "nose" of Mt. Stephen, dates to 1884. It is 200 m west of the Mt. Stephen Snowshed.

⓭ Field Yard The CPR lies between the community of Field and the Kicking Horse River. There are good viewpoints in Field and along the "backroad" to the west, which follows the 1884 railway grade. The author once saw 37 diesel-electric locomotives in the yard at one time. The Field station building dates to 1953. The only truly historic railway structures are the Telegraph Building, built in 1932, and the

older water tower, now a woodpecker haven. However, some of the CPR houses near the tracks are more than a century old. The signal just east of Field divides the Laggan and Mountain subdivisions. East of this signal, the movement of trains in the Laggan Subdivision is controlled from Calgary. West of this signal, the movement of trains in the Mountain Subdivision is controlled from Revelstoke. Thus, as far as the CPR is concerned, Field is in the Pacific time zone, while to everyone else, the community is in the Mountain time zone.

⓮ The Ottertail Diversion Completed in 1902, this diversion reduced the grade over Muskeg Summit west of Field. Highway 1 crosses the diversion at the rock cut 2.6 km west of town. You can also see the diversion from the first km of the Emerald Lake Road. See page 39.

⓯ Ottertail Bridge This 43-m bridge over the Ottertail River is visible briefly from Highway 1, 10.1 km west of Field. The earthworks of the original Ottertail Trestle Bridge are south of the highway bridge. See page 39.

⓰ The Black Bridge This 48-m truss bridge over the Kicking Horse River is visible from the highway shoulder, 14.5 km west of Field.

⓱ Mt. Hunter Trailhead Park at the Wapta Falls turnoff, 24.7 km west of Field. The Mt. Hunter trail is on the opposite side of Highway 1. It crosses the CPR in 225 m.

⓲ Leanchoil Siding Highway 1 parallels the CPR and the 106-car Leanchoil (lee-ANN-coil) Siding for almost 3 km near the west boundary of Yoho National Park. West of here, trains descend the 2 percent grade to Golden. Leth-na-Coyle was the Scottish home of CPR Vice-President Donald Smith's mother. The name was originally applied on the CPR to a station, now abandoned, 3.2 km east of the present siding. You can see the earthworks of the Leanchoil wye at the side of Highway 1, just west of the Wapta Falls Road.

Index

Author photo: Marnie Pole

The Author

Graeme Pole is an award-winning author and photographer who lives just down the Big Hill from the Spiral Tunnels, in Field, BC. He is a licensed interpretive guide, and a paramedic and Unit Chief with the British Columbia Ambulance Service.

The Spiral Tunnels and The Big Hill is his fifth book on the Canadian Rockies. The others are: *Canadian Rockies SuperGuide, The Canadian Rockies: A History in Photographs, Walks and Easy Hikes in the Canadian Rockies,* and *Classic Hikes in the Canadian Rockies.* All are published by Altitude Publishing.

With proceeds from the sales of his books, Graeme supports the reforestation work of Trees for the Future, 11306 Estona Dr., Box 1786, Silver Spring, MD, USA, 20915-1786.

Photography Sources and Credits